Eugen Scholz

Karl Fischer Titration

Determination of Water

Chemical Laboratory Practice

With 30 Figures and 8 Tables

Springer-Verlag
Berlin Heidelberg New York Tokyo 1984

Dr. Eugen Scholz
Production Manager Inorganic Chemicals
Riedel-de Haën AG
Wunstorfer Str. 40, D-3016 Seelze/Hannover
Federal Republic of Germany

Revised English translation of *Karl-Fischer-Titration* by E. Scholz in the series *Anleitungen für die chemische Laboratoriumspraxis,* Vol. 20, Springer-Verlag 1984
Edited by
F. L. Boschke, Heidelberg/FRG   W. Fresenius, Taunusstein/FRG
J. F. K. Huber, Wien/Austria   E. Pungor, Budapest/Hungary
G. A. Rechnitz, Newark/USA   W. Simon, Zürich/Switzerland
Th. S. West, Aberdeen/U. K.

ISBN 3-540-13734-3 Springer-Verlag Berlin Heidelberg New York Tokyo
ISBN 0-387-13734-3 Springer Verlag New York Heidelberg Berlin Tokyo

Library of Congress Cataloging in Publication Data. Scholz, Eugen. Karl Fischer titration (Chemical laboratory practice). Ref. translation of: Karl Fischer Titration. Bibliography: p. Includes index. 1. Karl Fischer technique. I. Title. II. Series. QD111. S41513 1984 544 84-14165 ISBN 0-387-13734-3 (U. S.)

This work is subject to copyright. All rights are reserved, whether the whole or part of the material is concerned, specifically those of translation, reprinting, re-use of illustration, broadcasting, reproduction by photocopying machine or similar means, and storage in data banks. Under § 54 of the German Copyright Law where copies are made for other than private use a fee is payable to "Verwertungsgesellschaft Wort", Munich.
© by Springer-Verlag Berlin Heidelberg 1984
Printed in Germany

The use of registered names, trademarks, etc. in this publication does not imply, even in the absence of a specific statement, that such names are exempt from the relevant protective laws and regulations and therefore free for general use.

Typesetting: SRZ Hartmann + Heenemann GmbH & Co. KG, Berlin
Printing: Brüder Hartmann GmbH & Co., Berlin   Bookbinding: Lüderitz & Bauer-GmbH, Berlin
2154/3020-543210

Preface

The Karl Fischer titration is used in many different ways following its publication in 1935 and further applications are continually being explored. At the present time we are experiencing another phase of expansion, as shown by the development of new titration equipment and new reagents. KF equipment increasingly incorporates microprocessors which enable the course of a titration to be programmed thus simplifying the titration. Coulometric titrators allow water determinations in the microgram-range: the KF titration has become a micro-method. The new pyridine-free reagents make its application significantly more pleasant and open up further possibilities on account of their accuracy.

To make the approach to Karl Fischer titrations easier, we have summarized the present knowledge in this monograph and we have complemented it with our own studies and practical experience. As this book should remain "readable", we have tried to keep the fundamentals to a minimum. Historical developments are only mentioned if they seem to be necessary for understanding the KF reaction. The applications are described more fully. Specific details which may interest a particular reader can be found in the original publications cited.

The referenced literature is in chronological order as the year of publication may also prove informative. Thus, [6902] for example denotes 69 for 1969 being the year of publication and 02 is a non-recurring progressive number. The referenced literature includes summaries which we hope will be of help to find the "right" publication easily.

Much assistance have been given in the preparation of this book for which the author is grateful. Aknowledgement is due to Riedel-de Haën AG, which made available all the facilities of an industrial concern, especially the literature, the laboratories and the office administration. Special thanks is rendered to Miss Hoffmann who carried out the laboratory work and to David M. Lee for his contribution to the translation of the English edition. Thanks is also due to a number of colleagues who assisted me in researching the literature. Further assistance shall be continued to be appreciated.

Seelze/Hannover, October 1984   Eugen Scholz

Contents

1	**Karl Fischer**	1
2	**The Karl Fischer Reaction**	3
2.1	Stoichiometry	3
2.1.1	The Effect of the Solvent	5
2.1.2	The Effect of the Water Concentration	5
2.2	Kinetics	6
2.3	Own Investigations	7
2.3.1	The Stoichiometry in Presence of Various Amines	7
2.3.2	The Alcohol-Sulphur Dioxide-Water System	10
2.4	The Karl Fischer Equation	13
3	**Titration Techniques**	15
3.1	Titrations Using a One-Component Reagent	15
3.2	Titrations Using a Two-Component Reagent	17
3.3	Coulometric Determination	19
3.4	Back Titration	21
3.5	Titrations at Different Temperatures	22
3.6	Continuous Determinations	23
3.7	Indication	24
3.7.1	Visual Indication	24
3.7.2	Instrumental Indication	25
4	**Reagent Solutions**	27
4.1	Reagent Solutions for Volumetric Titrations	27
4.2	Reagent Solutions for Coulometry	31
4.3	Stability of Solutions	34
4.4	Preparation of Solutions	38
4.5	Standardization of Titre	40
4.5.1	Standardization Using Solid Hydrates	41
4.5.2	Standardization Using Water	41
4.5.3	Stardardization Using Alcohol-Water-Mixtures	42

5	**Equipment**	45
5.1	Equipment for Volumetric Analysis	45
5.1.1	Simple Titration Set-Ups	45
5.1.2	Multipurpose Titration Equipment	48
5.1.3	Karl Fischer Titrators	49
5.2	Equipment for Coulometry	54

6	**Application and Sources of Error**	57
6.1	Working Medium	58
6.2	The pH	60
6.3	Side Reactions	62
6.4	Atmospheric Moisture	63
6.5	Indication	65
6.6	Sample Handling	66
6.6.1	Taking Samples and Sample Storage	67
6.6.2	Administration of Samples	67
6.6.3	Liquids	68
6.6.4	Solids	68
6.6.5	Gases	69
6.7	Safety Precautions	70

7	**Organic Compounds**	73
7.1	Hydrocarbons	73
7.2	Halogenated Hydrocarbons	74
7.3	Alcohols and Phenols	75
7.4	Ethers	76
7.5	Aldehydes and Ketones	76
7.6	Acids, Esters and Salts	78
7.7	Nitrogen-Containing Compounds	80
7.7.1	Amines and N-Heterocycles	80
7.7.2	Amides	82
7.7.3	Nitriles and Cyanohydrins	82
7.7.4	Hydrazine Derivatives	82
7.7.5	Other Nitrogen Compounds	82
7.8	Sulphur Compounds	82

8	**Inorganic Compounds**	85
8.1	Halides	85
8.2	Oxides, Hydroxides and Peroxides	87
8.3	Sulphur Compounds	88
8.4	Selenium and Tellurium Compounds	89
8.5	Nitrogen Compounds	90

8.6	Phosphorus Compounds	90
8.7	Arsenic and Antimony Compounds	91
8.8	Carbonates, Bicarbonates	91
8.9	Silicon Compounds	92
8.10	Boron Compounds	92

9	**Foodstuffs**	93
9.1	Handling Samples	95
9.2	Titration Techniques	96
9.3	Carbohydrates	97
9.3.1	Sugars and Sugar Products	97
9.3.2	Starch, Dextrins, Pectins, Flour	99
9.4	Fats	100
9.5	High-Protein Foods	101
9.6	Various Foodstuffs	101
9.6.1	Fruit Juices, Alcoholic Beverages	101
9.6.2	Mayonnaise	101
9.6.3	Baby Food	102
9.6.4	Coffee, Instant Coffee	102
9.6.5	Egg Powder	102
9.6.6	Pudding Powder	102
9.6.7	Cocoa Powder, Cocoa Beans	102
9.6.8	Chocolate	102
9.6.9	Almonds, Hazelnuts	102
9.6.10	Marzipan	103
9.6.11	Pastries	103
9.6.12	Pasta	103
9.6.13	Grain	103
9.6.14	Dried Vegetables, Dried Fruit	103
9.6.15	Meat and Meat Products	104

10	**Technical Products, Natural Products**	105
10.1	Technical Gases	105
10.2	Liquid Products of Mineral Oil	106
10.3	Insulating Oils, Lubricating Oils and Greases	107
10.4	Plastics	107
10.5	Ion Exchange Resins	108
10.6	Cellulose, Paper, Wood	108
10.7	Insulating Paper	108
10.8	Surfactants	109
10.9	Cigarette Smoke Condensate	109
10.10	Paints and Varnishes	109
10.11	Fertilizers	109

10.12	Cement	109
10.13	Minerals	110

Literature 111

Author Index 133

Subject Index 135

1 Karl Fischer

The Karl Fischer reaction is used in many laboratories throughout the world and the abbreviation "KF" is a term familiar to chemists. Certainly there are only a few names as well known in the field of analysis.

The man Karl Fischer is virtually unknown. The author endeavoured to obtain information and it was only Ernst Eberius, a friend of Karl Fischer since childhood and well known as the author of the German monograph "Wasserbestimmung mit Karl Fischer Lösung", who was able to help.

Karl Fischer was born on March 24th, 1901 in Pasing near Munich. He grew up in Leipzig where he attended school and university. In 1927 he joined a company in Berlin, later to be known as Edeleanu GmbH which belonged to the DEA-group. Karl Fischer became a petrochemist. He started up the Edeleanu plants for this company in many parts of the world, in Scotland and Sumatra, in Burma and Brazil and 12 plants in the USA alone. In 1936 he was appointed head of the company's laboratories in Berlin. In 1945 he went to the United States where he became professor of petrochemistry at the University of Maryland. In 1950 he returned to Germany, rejoining DEA as head of the company's central laboratory in Hamburg. He died of a myocardial infarction on April 16th, 1958.

His friend Ernst Eberius described Karl Fischer as an imaginative person with an intuitive mind who was always able to entertain. His large pipe collection which included some of the most unusual specimens from all parts of the world, was quite famous. Appraised were his spirits which were produced in 10 litre containers especially for him. He was famous for his hobby – conjuring. He was able to perform an unending succession of conjuring tricks without even his closest friends perceiving how. He was constantly endeavouring to complement his repertoire, most of his ideas originating from Java.

In his professional life he was just as versatile, imaginative and witty. As a schoolboy he had built a microscope from parts of an artillery shell and made slide preparations using a self-made microtome. In this way he supplemented his pocket-money. This wealth of ideas also characterized his later activities as a petrochemist which were very successfull indeed. His patents and publications are evidence that he did not restrict his capabilities to his specialized field: deparaffination of petroleum, separation of unsaturated compounds, analysis of paraffin, conditioning of drilling muds, aluminium silicate cracking catalysts, determination of aromatics in white oils and aids for microscopy. He was particularly proud of the alginate impregnation of jute sacks he had introduced.

His water reagent was also a by-product of the petrochemistry. The plant for the Edeleanu process tended to corrode, caused by the water content in the sulphur dioxide. Thus, a quick method for determining water was required, which was developed in a few days only – told Ernst Eberius. Fischer resorted to the well-known Bunsen reaction. He found it applicable to this particular problem if the liberated acid were neutralized. To this he used pyridine – "it happened to be on the shelf" – told Eberius.

Although well-meaning colleagues warned him against publishing his "in different aspects muddy concoction" he published in 1935 his "Neues Verfahren zur maßanalytischen Bestimmung des Wassergehaltes in Flüssigkeiten und festen Körpern".

Thus, the "conjuring trick" of Karl Fischer came about which has since kept two generations of chemists occupied without their managing to completely unravel the secrets of this "trick".

2 The Karl Fischer Reaction

2.1 Stoichiometry

To develop his method for the determination of water, Karl Fischer [3501] resorted to the well-known Bunsen reaction which is used to determine sulphur dioxide in aqueous solutions:

$$2\,H_2O + SO_2 + I_2 \rightarrow H_2SO_4 + 2\,HI. \qquad (2.1)$$

This reaction seemed to be applicable to the determination of water in non-aqueous solutions if an excess of sulphur dioxide was present and if the liberated acid was neutralized by an organic base. Fischer used "insbesondere Anilin und Pyridin" (aniline and pyridine, in particular). Thus the Karl Fischer reagent was composed: a solution of sulphur dioxide and iodine in pyridine and methanol. Fischer postulated the reaction to proceed as follows:

$$2\,H_2O + SO_2 \cdot (C_5H_5N)_2 + I_2 + 2\,C_5H_5N$$
$$\rightarrow (C_5H_5N)_2 \cdot H_2SO_4 + 2\,C_5H_5N \cdot HI. \qquad (2.2)$$

It is not clear from Fischer's publication whether he formulated the reaction as a verification or as an extension of the Bunsen reaction (2.1). Today we know this equation to be incorrect. Equation (2.2) infers a mole ratio

$$H_2O : I_2 : SO_2 : Py = 2 : 1 : 1 : 4 \qquad (2.3)$$

(Py = Pyridine.)

This mole ratio was investigated a few years later by Smith, Bryant and Mitchell [3901]. They changed the composition step by step until the water determination was no longer possible. It was similarily shown that methanol takes part in the reaction and does not act solely as a solvent. They found the mole ratio to be:

$$H_2O : I_2 : SO_2 : Py : CH_3OH = 1 : 1 : 1 : 3 : 1 \qquad (2.4)$$

They suggested that the reaction takes place in two distinct steps (2.5) and (2.6):

$$I_2 + SO_2 + 3\,Py + H_2O \rightarrow 2\,Py \cdot HI + Py \cdot SO_3, \qquad (2.5)$$

$$Py \cdot SO_3 + CH_3OH \rightarrow Py \cdot HSO_4CH_3. \qquad (2.6)$$

First an intermediate, pyridine-sulphur trioxide is formed, which is solvolyzed by methanol. Using a methanol-free reagent, this intermediate reacts with a second molecule of water:

$$Py \cdot SO_3 + H_2O \rightarrow Py \cdot H_2SO_4 \tag{2.7}$$

If the reactions (2.5) and (2.7) are combined, an equation for non-methanolic solutions is obtained which is identical with the Fischer equation (2.2).

The equation for methanolic reagents was modified a few years later including the formulation of iodine and sulphur dioxide as pyridine adducts [Aqu. p. 71].

$$H_2O + Py \cdot I_2 + Py \cdot SO_2 + Py + CH_3OH \rightarrow 2\,Py \cdot HI + Py \cdot SO_3, \tag{2.8}$$

$$Py \cdot SO_3 + CH_3OH \rightarrow Py \cdot HSO_3CH_3. \tag{2.9}$$

Especially since the publication of the monograph "Aquametry" by Mitchell and Smith, these presentations of the reaction have been generally accepted, and they are still used in many textbooks on analytical chemistry today.

But there are a few doubts. The hypothetical intermediate, the pyridine/sulphur trioxide, could be synthesized by pouring a mixture of pyridine, sulphur dioxide and iodine onto ice [3901]. A water reagent is produced in an aqueous medium! And, surprisingly, it reacts very slowly with cold water and methanol. The existence of the pyridine/iodine adduct was questioned by Blasius and Wachtel [5302]. They found this substance to be present only in small amounts in methanolic solutions.

Further systematic investigations on stoichiometry are not known. But a few kinetic studies contribute to questions of stoichiometry. Cedergren [7402] found that the reaction is first-order with respect to water, iodine and sulphur dioxide. This means that these three components react in a mole ratio of 1:1:1.

This result was confirmed by Verhoef and Barendrecht [7602]. Furthermore, they also stated that pyridine does not play an active role in the KF reaction. Other solutions buffered to the same pH using dichloroacetate or salicylate showed the same reaction rates. Pyridine is not an important reactant in the KF solution. It merely acts as a buffer.

The aforementioned authors investigated the dependance of the reaction rate on the pH and stated that it is not sulphur dioxide but a sulphurous base which is oxidized. Sulphur dioxide reacts with the methanol according to Eq. (2.10).

$$2\,CH_3OH + SO_2 \rightleftharpoons CH_3OH_2^+ + SO_3CH_3^-. \tag{2.10}$$

The addition of bases to the methanolic solution shifts this equilibrium (2.10) predominantly to the right:

$$CH_3OH + SO_2 + RN \rightarrow [RNH]SO_3CH_3 \tag{2.11}$$

(RN = Base.)

2.1 Stoichiometry

Based on these studies the Karl Fischer reaction can be formulated as follows:

$$H_2O + I_2 + [RNH]SO_3CH_3 + 2\,RN \rightarrow [RNH]SO_4CH_3 + 2[RNH]I. \tag{2.12}$$

It should be pointed out that this reaction is restricted to methanolic solutions.

2.1.1 The Effect of the Solvent

Methanol is the solvent normally used for Karl Fischer titrations and Eq. (2.9) includes methanol as a reactant. But non-methanolic reagents are mentioned in a few publications. Benzene-based reagents have been reported by Fischer [3501] and also by Smith, Bryant and Mitchell [3901]. Their application however proved unsuccessful in practice. A deviation in the stoichiometry was mentioned but neither tested nor proven. We also tried to titrate water in a number of organic solvents using a KF reagent based on 2-methoxyethanol. The recovery rates for water in toluene, xylene, ethyl acetate, chloroform, formamide, acetone and dimethyl sulphoxide were found to be too low. Another stoichiometric relation could not be deduced.

Klimova, Sherman and L'vov [6703, 6704] used reagents containing dimethyl formamide instead of methanol. The titre of such reagents depends on the medium in which the titration is carried out. When using dimethyl formamide for the sample, the titre is double that when using methanol. They assume a mole ratio $H_2O:I_2$ of 1:1 in methanol and a reaction following Eq. (2.12), and in dimethyl formamide a mole ratio of 2:1 a reaction following Eq. (2.2).

The interrelation between water equivalent and working medium (i.e. the solvent for the sample) was extensively covered by Sherman [8006]. He titrated water in different solvents using the above reagent based on DMF. These investigations also show that in aprotic solvents the titre is approximately double of that in methanol. Intermediate values, as obtained for propanol, butanol and other long chain alcohols would appear to be abnormalous.

In summarizing, it can be said that the stoichiometry of the KF reaction can change in certain non-methanolic solvents, but quantitative correlations cannot be substantiated by present investigations.

2.1.2 The Effect of the Water Concentration

The stoichiometry of the KF reaction is also influenced by the concentration of water in the working medium. Bonauguri and Seniga [5502] found that the titration of undiluted water required less KF reagent than the more usual method of titrating water diluted with methanol. Verhoef [7703] found the stoichiometry to change during coulometric determinations if the water content exceeds 1 mol/l. He assumed sulphurous acid to be produced at higher concentrations:

$$SO_2 + H_2O \rightarrow HSO_3^- + H^+. \tag{2.13}$$

This mode of reaction resembles the Bunsen reaction (2.2) for aqueous media.

2.2 Kinetics

There are only a few publications dealing with the kinetics of the Karl Fischer reaction and our knowledge is limited.

Cedergren [7404] investigated the course of the reaction in a coulometric cell. He found the rate of reaction to increase with increasing concentrations of sulphur dioxide, iodine and water. The reaction is of first order with respect to each of these three components. He found a value for K, the reaction rate constant, as defined by Eq. (2.14) of:

$K = (1.2 \pm 0.2) \cdot 10^3 \cdot l^2 \cdot mol^{-2} \cdot s^{-1}$,

$\log K = 3.08 \pm 0.08$.

$$-d[I_2]/dt = K[I_2] \cdot [SO_2] \cdot [H_2O] \tag{2.14}$$

The rate seemed to be independent of the pyridine concentration provided it was present in sufficient quantities.

Verhoef and Barendrecht investigated the reaction under similar conditions [7602] also using a rotating ring-disk electrode [7703]. They could confirm the results of Cedergren.

The former authors also stated the reaction rate constant K to depend on the pH of the solution (Fig. 2–1). In acid solutions up to pH 5, the logarithm of the reaction rate constant increases linearly with the pH. In the range from pH 5.5–8, the reaction rate has a constant value, $\log K = 3.12$. Beyond pH 8.5 the reaction rate again increases with the pH, but only slightly and presumedly because of a side reaction.

The shape of the log K vs. pH curve suggests that it is not sulphur dioxide itself, but rather a sulphurous base which is oxidized. Titrating a methanolic solution of sulphur dioxide with TMAH they found only one inflection point. The solution contains a moderately weak monobasic acid. They deduced methyl sulphurous acid to be produced according to Eq. (2.10). From the titration curve, an apparent dissociation constant for the reaction was calculated as:

$$K_a = [CH_3OH_2^+] \cdot [SO_3CH_3^-]/[SO_2]. \tag{2.15}$$

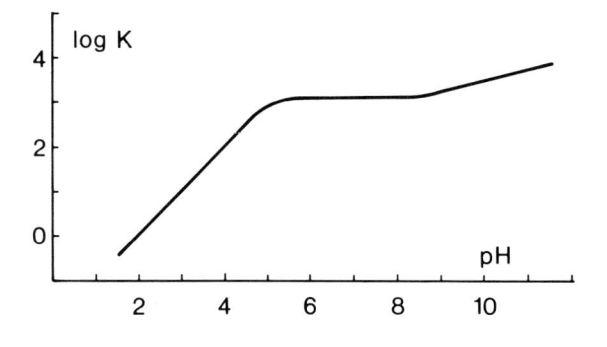

Fig. 2–1 Dependance of the reaction rate constant K on the pH according Verhoef and Barendrecht [7602]

2.3 Own Investigations

They found a value of 6.02 for pK_a, independent of the water concentration. This is a considerable prerequisite for the stoichiometry of the KF reaction.

Cedergren's findings, that the pyridine concentration does not influence the reaction rate, could also be confirmed. Other buffers of the same pH gave the same reaction rate. Pyridine is not a reactant in the KF reaction as pointed out in Eqs. (2.8) and (2.9), it merely acts as a buffer.

The rate of reaction is greatly affected by the concentration of iodide. Iodine reacts with the iodide, to yield triiodide:

$$I_2 + I^- \rightarrow I_3^- \tag{2.16}$$

This equilibrium tends to the right and triiodide is very stable in methanolic solutions. Both compounds react with methyl sulphurous acid at different reaction rates. They found the following values [7602].

$$K_{I_3} = 3.46 \cdot 10^2,$$
$$K_{I_2} = 8.8 \cdot 10^6.$$

The large difference in the reaction rate of iodine and triiodide is remarkable and Verhoef and Barendrecht considered the following reaction intermediates:

$$I^{\delta-} - I^{\delta+} \ldots \left[O - \overset{O}{\underset{\|}{S}} - OCH_3 \right]^- \quad \text{or} \tag{2.17}$$

$$\left[O - \overset{O}{\underset{\|}{S}} - OCH_3 \right]^- \tag{2.18}$$

These intermediates are then hydrolyzed by water to form hydroiodic acid and monomethyl sulphuric acid.

2.3 Own Investigations

2.3.1 The Stoichiometry in Presence of Various Amines

Introducing the new pyridine-free reagents [8002, 8003], we had to investigate whether the reaction follows the Karl Fischer equation according to Eqs. (2.5) and (2.6). We used the acid-base titration in non-aqueous solution as a suitable tool to follow the KF reaction, because several reactants of the Karl Fischer solution have acid-base properties.

If a methanolic solution containing 1.5 mmol of diethanolamine is titrated with perchloric acid, the content of the diethanolamine can be determined quantitatively (Fig. 2–3, curve A). When titrating the same solution with TMAH, no inflection point is indicated; the solution does not contain any acid (Fig. 2–2, curve A).

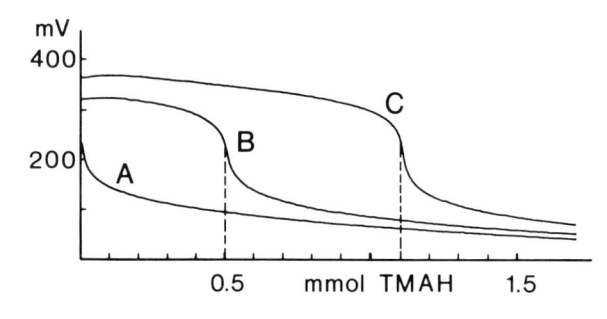

Fig. 2-2 The KF reaction in the presence of diethanolamine. Titration of intermediates with TMAH (0.1 mol/l in 2-propanol). A: 20 ml methanol + 1.5 mmol diethanolamine. B: as in A + 0.5 mmol SO$_2$. C: as in B + 0.3 mmol H$_2$O + 0.3 mmol I$_2$

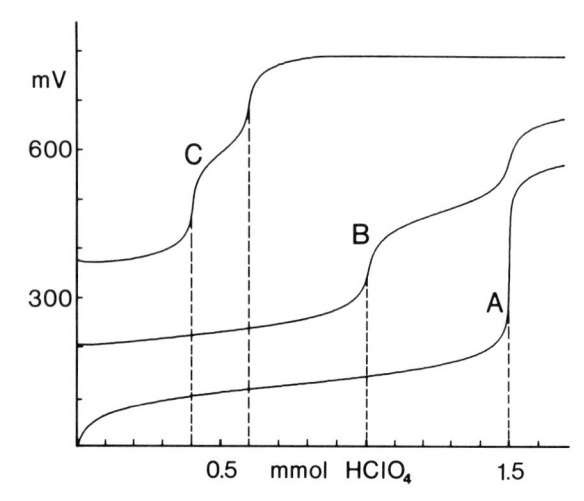

Fig. 2-3 The KF reaction in the presence of diethanolamine. Titration of intermediates with perchloric acid (0.1 mol/l in 2-propanol). A, B, C as in Fig. 2-2. (Each ordinate displaced by 100 mV)

If the methanolic solution contains 1.5 mmol of diethanolamine and 0.5 mmol of sulphur dioxide, two end points are obtained upon titration with perchloric acid: 1.0 mmol of diethanolamine and 0.5 mmol of a weak base (Fig. 2-3, curve B). By titrating the same solution with TMAH we found 0.5 mmol of a weak acid (Fig. 2-2, curve B). 0.5 mmol of sulphur dioxide had produced 0.5 mmol of a weak acid and 0.5 mmol of a weak base.

Assuming the reaction of Eq. (2.11) it can be deduced that the sulphur dioxide produces the diethanolammonium salt of the methyl sulphurous acid.

Titrating with TMAH the diethanolammonium salt reacts as a weak acid according to Eq. (2.19).

$$[(HOCH_2CH_2)_2NH_2]SO_3CH_3 + [(CH_3)_4N]OH$$
$$\rightarrow [(CH_3)_4N]SO_3CH_3 + (HOCH_2CH_2)_2NH + H_2O \qquad (2.19)$$

Titrating with perchloric acid the methyl sulphite ion reacts as a weak base according to Eq. (2.20).

$$[(HOCH_2CH_2)_2NH_2]SO_3CH_3 + HClO_4$$
$$\rightarrow [(HOCH_2CH_2)_2NH_2]ClO_4 + HSO_3CH_3. \qquad (2.20)$$

2.3 Own Investigations

These titrations indicate that the diethanolammonium salt of the methyl sulphurous acid is actually present in the KF solution prior to KF titration.

Then we oxidized the methanolic solution (1.5 mmol of diethanolamine and 0.5 mmol of sulphur dioxide) by adding 0.3 mmol of iodine and 0.3 mmol of water – i.e. we performed the Karl Fischer reaction. By titrating this solution with TMAH, we found 1.1 mmol of a weak acid, 0.6 mmol more than prior to oxidation (Fig. 2–2, curve C). 0.3 mmol of iodine and 0.3 mmol of water had produced 0.6 mmol of diethanolammonium salt. It could be the diethanolammonium salt of the hydroiodic acid according to Eq. (2.12). If the oxidized solution is titrated with perchloric acid (Fig. 2–3, curve C), only 0.4 mmol of free diethanolamine and 0.2 mmol of methyl sulphite are detectable. 0.3 mmol of methyl sulphite had been oxidized to methyl sulphate according to Eq. (2.12). 0.9 mmol of diethanolamine cannot be detected. It had been neutralized by strong acids: i. e. 0.3 mmol of methyl sulphuric acid and 0.6 mmol of hydroiodic acid.

These investigations confirm the stoichiometry expressed by Eqs. (2.11) and (2.12).

In pyridine-based Karl Fischer solutions, the stoichiometry can be proven by similar means. Figures 2–4 and 2–5 indicate the same molar ratio between the KF reactants according to Eq. (2.12). Differences gradually become apparent with acidimetric and alkalinimetric titrations. The potential drops are greater for titrations with

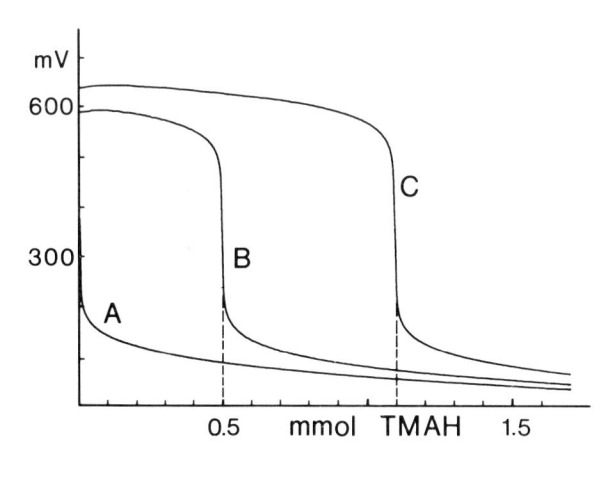

Fig. 2-4 The KF reaction in the presence of pyridine. Titration of intermediates with TMAH (0.1 mol/l in 2-propanol). A: 20 ml methanol + 1.5 mmol pyridine. B: as in A + 0.5 mmol SO$_2$. C: as in B + 0.3 mmol H$_2$O + 0.3 mmol I$_2$

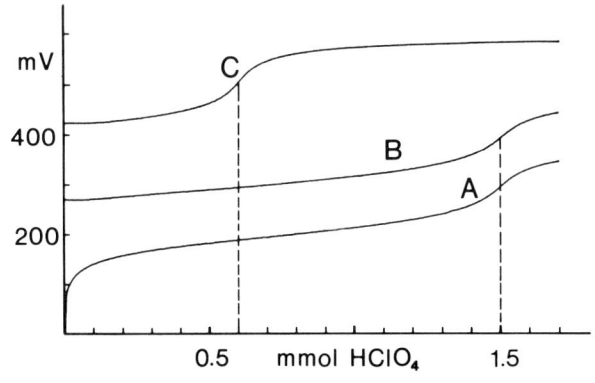

Fig. 2-5 The KF reaction in the presence of pyridine. Titration of intermediates with perchloric acid (0.1 mol/l in 2-propanol). A, B, C as in fig. 2–4. (Each ordinate displaced by 100 mV)

TMAH because of the weaker basicity of pyridine, i.e. stronger acidity of pyridinium methyl sulphite. For the same reason the potential drops are lesser for titrations with perchloric acid. Pyridine and the methyl sulphite ion can not be detected separately. Both compounds have approximately the same basicity. Thus, the methyl sulphurous acid can not be neutralized completely by pyridine as supposed by Eq. (2.11). The concentration of the methyl sulphite ion is lower and the KF reaction proceeds more slowly as indicated by Verhoef and Barendrecht [7602].

Sodium acetate is also recommended as a Karl Fischer base [5301, 7704]. When titrating these solutions with perchloric acid or TMAH, the graphs resemble those obtained with diethanolamine. Imidazole [8203] is similar to pyridine although the potential drops are greater for acidimetric titration. The graphs show a slight point of inflection for the methyl sulphite anion. In accordance with Eq. (2.11) it is apparent that the slightly higher basicity of imidazole, as compared to pyridine, is sufficient to neutralize the methyl sulphurous acid. The Karl Fischer reaction is accelerated using imidazole as a base. In both cases, sodium acetate and imidazole, the graphs confirm the molar ratio according Eq. (2.12).

2.3.2 The Alcohol-Sulphur Dioxide-Water System

The esterification of sulphur dioxide with alcohol according Eq. (2.10) appears to be a considerable prerequisite for the course of the KF reaction. On the other hand, sulphur dioxide could react with water producing sulphurous acid according Eq. (2.13) as a reaction competitive to esterification. Considering the two reactions together, the following equilibrium in the alcohol/sulphur dioxide/water system seems to be plausible:

$$
\begin{array}{c}
A \\
H_2O + SO_2 + ROH \\
\swarrow \qquad\qquad\qquad \searrow \\
H^+ + SO_3H^- + ROH \rightleftharpoons H_2O + SO_3R^- + H^+ \\
C \qquad\qquad\qquad\qquad B
\end{array}
\qquad (2.21)
$$

From the original components, A, the state B could be formed by an active alcohol in the absence of water. If no alcohol is present, excessive water could form state C. If both alcohol and water are present, the equilibrium B–C is possible. State A could be attained when the amounts of water and alcohol are low, i.e. when working in an aprotic solvent medium.

The equilibrium A–B–C could predestinate a subsequent oxidation by iodine. The KF reaction depends on the solvent. We attempted to confirm this dependance by investigating different solutions using acid-base-titration.

The equilibrium B–C could depend on the reactivity of the alcohol. When solutions of sulphur dioxide in various alcohols are titrated with TMAH the different reactivities become evident (Fig. 2–6). A methanolic solution shows only one inflection point (curve A). This agrees with Verhoef and Barendrecht [7602] that only methyl sulphurous acid is present, thus corresponding to the equilibrium state B. In 1-propanol,

2.3 Own Investigations

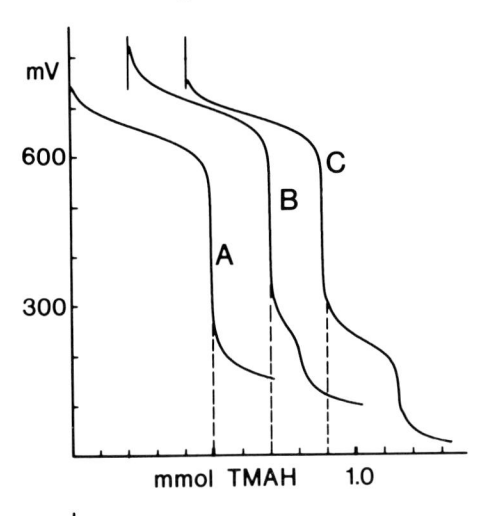

Fig. 2-6 Titration of sulphur dioxide in different alcohols with TMAH. 20 ml alcohol (A: methanol, B: 1-propanol, C: 2-propanol) + 0.5 g water + 0.5 mmol sulphur dioxide are titrated with TMAH (0.4 mol/l in 2-propanol). (Each abscissus displaced by 0.2 mmol)

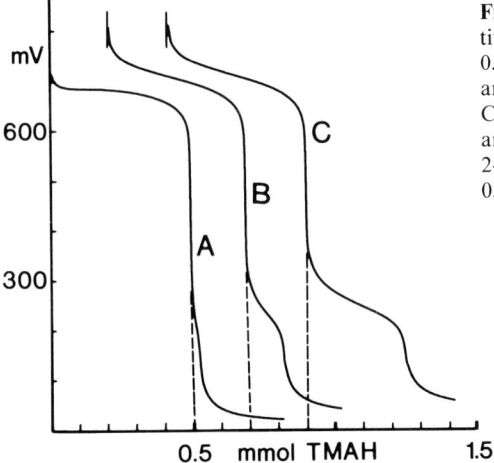

Fig. 2-7 The effect of water on the titration of sulphur dioxide in 2-propanol. 0.5 mmol sulphur dioxide and differing amounts of water (A: 10 mmol, B: 20 mmol, C: 40 mmol), are added to 20 ml 2-propanol and titrated with TMAH (0.4 mol/l in 2-propanol). (Each abscissus displaced by 0.2 mmol)

however, a second equivalence point is indicated (curve B), due to the second level of sulphurous acid. In this solvent only 80% appears as ester, the remaining 20% is as sulphurous acid. In 2-propanol (curve C) the equilibrium is shifted even further towards state C and approximately 50% is present as sulphurous acid. These three titrations confirm that the equilibrium B–C depends on the nature of the alcohol.

From Eq. (2.21) it would be expected that the concentration of water also influences the equilibrium B–C. Figure 2–7 confirms this. In 2-propanol, sulphur dioxide is present as sulphurous acid to an extent of 6% at a water concentration of approximately 0.5 mol/l, 20% at 1 mol/l and 50% at a concentration of 2 mol/l.

Based on these titration results the equilibrium from Eq. (2.21) ist not improbable. However, if this equilibrium exists, it should also influence the following iodometric oxidation which essentially is the KF reaction. If state B is oxidized, the reaction follows Eq. (2.12), whereby the $H_2O:I_2$ mole ratio is 1:1. This is the Karl Fischer reaction. If state C is oxidized, the reaction follows Eq. (2.2) and the $H_2O:I_2$ ratio is 2:1. This is the Bunsen reaction. This means that the stoichiometry of the KF reaction

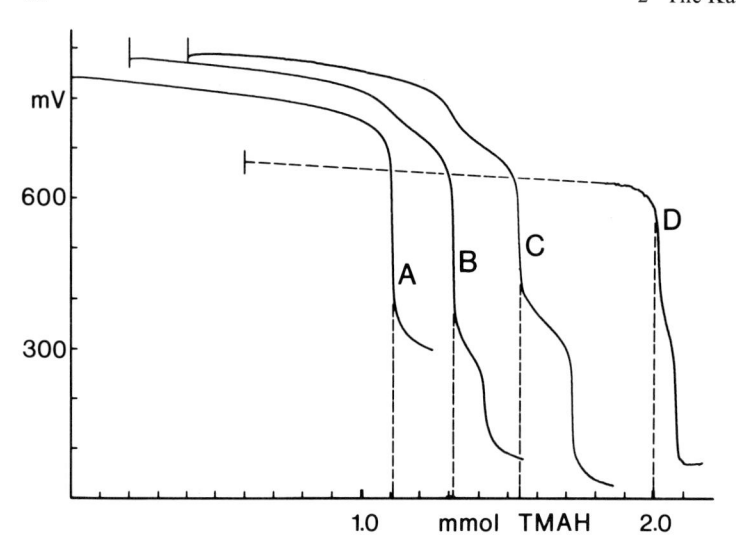

Fig. 2-8 The KF reaction in different solvents. 20 ml of the solvent (A: methanol, B: 2-propanol, C: tert.-butanol, D: pyridine) + 1.5 mmol pyridine + 25 mmol water + 0.5 mmol SO$_2$ + 0.3 mmol I$_2$ are mixed. The mixture is titrated with TMAH (0.4 mol/l in 2-propanol). (Each abscissus displaced by 0.2 mmol)

could be changed if unsuitable alcohols are used or if the concentration of water is high.

As the KF reaction and the Bunsen reaction produce different amounts of acid, they can be distinguished from each other by acidimetric and alkalinimetric titration. The KF reaction produces 2 moles of acid per mole iodine, the Bunsen reaction produces 3 moles (3 equivalents). This difference can be determined by titration with TMAH. Figure 2–8 depicts the results. In methanol 2 moles of acid are produced. The KF reaction proceeds stoichiometrically. In 2-propanol 2.05 moles of acid are liberated, in tertiary butanol 2.12 moles and in pyridine 3.00 moles. The Bunsen reaction proportion increases from approximately 5% in 2-propanol to 12% in tert-butanol and nearly 100% in pyridine.

These results can be confirmed by acidimetric titration. The Karl Fischer reaction produces 3 equivalents of strong acid (alkyl sulphuric acid and hydroiodic acid). The Bunsen reaction produces 4 equivalents of strong acid (sulphuric acid and hydroiodic acid). These acids are neutralized by the amine present in the solution. By titrating with perchloric acid, the excess of amine is detected and the acid produced can be calculated. Figure 2–9 shows the titrations. In methanol 3.0 equivalents of acid are produced, in 2-propanol 3.0, in tert-butanol 3.15 and in acetone 3.5 equivalents.

Both titrations indicate that in methanol the oxidation reaction proceeds correspondingly to the KF reaction. Here, the state B of the equilibrium in Eq. (2.21) would appear to actually exist. In 2-propanol and tert-butanol, the Bunsen reaction is also evident to a certain degree. Here, an equilibrium state between B and C is oxidized. But the influence of the Bunsen reaction is significantly less than expected by the content of sulphurous acid as determined in Figs. 2–5 and 2–6. This would be plausible

2.4 The Karl Fischer Equation

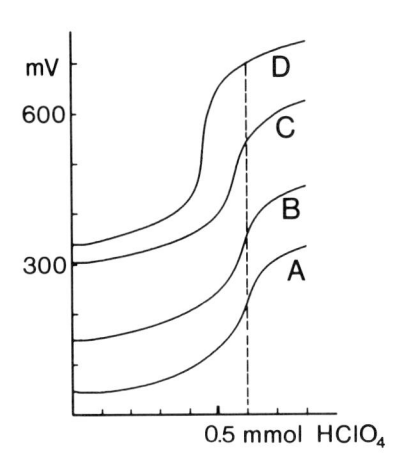

Fig. 2-9 The KF reaction in different solvents. 20 ml of the solvent (A: methanol, B: 2-propanol, C: tert.-butanol, D: acetone) + 1.5 mmol imidazole + 25 mmol water + 0.5 mmol SO$_2$ + 0.3 mmol I$_2$ are mixed. The mixture is titrated with perchloric acid (0.4 mol/l in 2-propanol). (Each ordinate displaced by 100 mV)

if the oxidation of state B were kinetically favourable to that of state C or if the equilibrium B–C were shifted during oxidation.

Pyridine and acetone behave quite differently as solvents for the KF titration. In pyridine the oxidation follows the Bunsen reaction (2.2) quantitatively. In acetone the iodometric oxidation is retarded as the iodine colour decreases slowly. As the titration proceeds, the reaction is accelerated, possibly due to the propanol added by the titrant. The amount of acid found is not reproducible. This is not typical for neither the KF reaction nor the Bunsen reaction and we suppose the state A to exist in acetone.

2.4 The Karl Fischer Equation

Summarizing the results of all investigations we can formulate a few important statements:

The reaction of sulphur dioxide with the alcohol producing a monoalkyl ester of the sulphurous acid is a basic requirement for the Karl Fischer reaction. Thus, the Bunsen reaction is prevented and the stoichiometric course of the KF reaction is ensured. The rate of reaction depends on the pH of the working medium, preferably on the amine applied. The monoalkyl ester of the sulphurous has to be neutralized, as the alkyl sulphite anion actually reacts in the Karl Fischer solution. The practical implications from these statements are discussed in Sections 6.1 and 6.2.

Based on our present knowledge the Karl Fischer reaction must be formulated as follows:

$$ROH + SO_2 + R'N \rightleftharpoons [R'NH]SO_3R,$$
$$H_2O + I_2 + [R'NH]SO_3R + 2R'N \rightarrow [R'NH]SO_4R + 2[R'NH]I. \quad (2.22)$$

In this formulation ROH is a "reactive" alcohol and R'N a "suitable" base.

For non-alcoholic solutions like formamide [6301], alkyl carbonate [8005] or pyridine no applicable equation can be formulated today.

3 Titration Techniques

Karl Fischer had originally recommended to titrate the solvent under investigation directly with the KF solution. In the following years, this procedure was further developed leading to an improvement in accuracy and reliability and to an extended applicability. The development of titration equipment led to modified techniques, too. This has resulted in three present-day standard techniques with specific advantages dependig on their field of application.

1. Volumetric titration using a one-component reagent.
2. Volumetric titration using a two-component reagent.
3. Coulometric determination.

These three techniques and several modifications are preferably used today and they are described in this section.

3.1 Titrations Using a One-Component Reagent

The one-component reagent contains all the substances necessary for the KF reaction – iodine, sulphur dioxide, pyridine and methanol – in one solution. It was introduced in this form by Karl Fischer. In the decades to follow, the composition was modified, the concentrations were changed, the methanol was replaced partly or completely by other solvents, and bases other than the unpleasant pyridine have been employed (see Section 4).

The one-component solution has become the most preferred form of the KF reagent today and is referred to in many analytical handbooks, for example in the European Pharmacopoeia, United States Pharmacopoeia, ASTM E 203 [8112], ACS "Reagent Chemicals" and ISO 760 [7806]. To a large extent, the working techniques do not depend on the specific composition of the KF reagent used. The following method has proved very practical, regardless of whether the titration is performed manually or using an automatic titrator, and is not specific to either visual or instrumental end-point indication.

Originally, the water content of a liquid sample had been determined titrating it directly with the KF reagent. As this procedure includes a few sources of error, it has been modified. Now, methanol is generally used as the working medium. It is dried by a pre-titration with the KF reagent first. Then the sample is added and titrated in the same manner. This procedure ensures a reliable determination of water.

The titration vessel, depending on its size, is filled with 10–200 ml methanol. The methanol is then pre-titrated to a stable end point with the KF reagent. This pre-titration is extremely important as the methanol is never absolutely moisture-free and especially as it is very hygroscopic. The titration vessel itself can still contain adherent moisture, even if taken directly out of a drying oven. Muroi and Ogawa [6302] found up to 5 mg residual water. The humidity in the air must also be taken in account. 1 Litre of air can contain as much as 20 mg of water and the titration vessel, depending on its volume, a corresponding amount. All these sources of errors can be eliminated by introducing the methanol into the titration vessel and titrating it to complete dryness with the KF solution prior to the investigation of the sample. The end-point stability serves as an indicator for a good desiccation. A dehydrated titration cell will remain stable for several minutes without reversal of the end point.

The dried methanol serves as the working medium and as the solvent for the sample to be analysed. The sample is added to the pre-dried methanol and dissolved if possible. The water contained is titrated immediately in order to reduce the errors caused by atmospheric moisture. The titration is performed in the same way as the pre-titration, whereby the same stable end point is desirable. This can be achieved if the substance is completely dissolved (or remains completely insoluble) and no side reaction occur.

Carrying out a Karl Fischer titration, the analyst observes an outstanding peculiarity: the delayed end point adjustment. Adding the first few drops of KF reagent to the working medium it becomes brown in colour. The end point seems to have been reached. However, the brown colour vanishes after a few seconds and the titration must be continued. As the real end point is approached, the brown colour persists longer and longer and the titration becomes more and more dragging. If the brown colour persists longer than 10–20 s the titration is regarded as finished. By using an automatic titrator with instrumental indication, the same behaviour is found. Figure 3–2, curve C, represents the pre-titration of 20 ml of methanol. It takes about 6 min. The following titration of 40 mg of water, as shown in Fig. 3–1, curve C, takes 10 min. The titration time is influenced by the KF reagent, the instrument used, the amount of water to be analysed, and the experience of the technician. Details will be discussed in the following sections.

The water content of the sample is calculated from the amount of KF solution consumed and the titre of the solution. (See Section 4.5.)

In practice, this standard procedure is modified in different ways. Earlier, methanol was not used as a working medium at all and organic liquids had been titrated directly. As this can affect the stoichiometry of the KF reaction, an alcoholic working medium should be employed on principle. (See Section 2.1.1.)

Even then, when an alcohol-based working medium is used, pre-titration is sometimes not carried out. Its water content is eliminated by means of a "blank titration". Despite the fact that two titrations are necessary, it is doubtful whether this "blank titration" constitutes the same set of interferences as in the main titration. It is certain that these sources of errors are eliminated by a pre-titration.

In many cases it is advisable to replace methanol by other solvents. When aromatics are being determined, propanol should be used as the working medium. For fats, a mixture of chloroform and methanol is better. Occasionally, non-alcoholic solvents are used. Pyridine is used, for example when ketones are being determined in order to inhibit the formation of ketals with alcohol. In many cases, the choice of a suitable working medium (see Section 6.1) is the most important decision to ensure the best possible water determination.

"Successive titrations" are a variation of the KF titration. After titrating the first sample, the next sample is added to the same solution. The titrated solution thus acts as the dehydrated working medium for the next determination. The amount of work involved is thus reduced, as the otherwise necessary pre-titration is no longer required. The KF reaction is also accelerated due to the accumulation of sulphur dioxide and amine. The "successive titration" is largely restricted to alcohol samples because otherwise the proceeding addition of samples would reduce the alcohol concentration to a level which would alter the stoichiometry. As a general rule it can be said that the alcohol content in the titration vessel should not fall below 50%. The limit depends on the alcohol used. For methanol, a content of 30% will suffice. A prerequisite for this "successive titration" is of course that only those substances are being analysed which do not react with one another.

3.2 Titrations Using a Two-Component Reagent

Methanolic one-component reagents have an unsatisfactory shelf life, which means they cannot be manufactured or distributed commercially. Thus, reagent manufacturers offered a product comprising of two components. Component A is a solution of sulphur dioxide and pyridine in methanol, component B is a solution of iodine in methanol. Both solutions are combined prior to use by the analyst, thus constituting a one-component reagent.

Johansson [4701] suggested to use the two solutions separately, i.e. to use component A (solvent component) as a solvent and to titrate with component B (titrant component). In this way both components are combined in the titration vessel. Although this method has distinct advantages, it did not became popular for a long time. The introduction of pyridine-free reagents – Reaquant by Baker in 1979 and Hydranal by Riedel-de Haën in 1980 – advanced this working technique.

To titrate water using a two-component reagent, component B (titrant component) is filled into a burette. Component A (solvent component) acts as the solvent for the sample. It is placed into the titration vessel and dried by pre-titration with component B to a stable end point. Then the sample is placed into the titration vessel and the water contained is titrated with the titrant component. The water content is calculated from the volume of titrant and from its titre.

The two-component technique has distinct advantages [8002]: using the solvent component as the working medium ensures a high excess of sulphur dioxide and amine. This accelerates the KF reaction and reduces the titration time. Figure 3–1 shows the course of the titration of 40 mg of water with three commercially available reagents. The titration with the two-component reagent (curve A) takes less than two minutes. The advantages become more significant when small amounts of water are determined. This is evident from Fig. 3–2, which shows the titration of 2 mg of water.

Using a one-component reagent for the titration of small amounts of water, only a few drops of reagent solution are needed. The sulphur dioxide and amine contents in the working medium are very low, and the equilibrium is reached only very slowly and thus the end point is in doubt. The same can be said for the pre-dehydration of the working medium, as this is also a titration of a small amount of water. If this pre-titra-

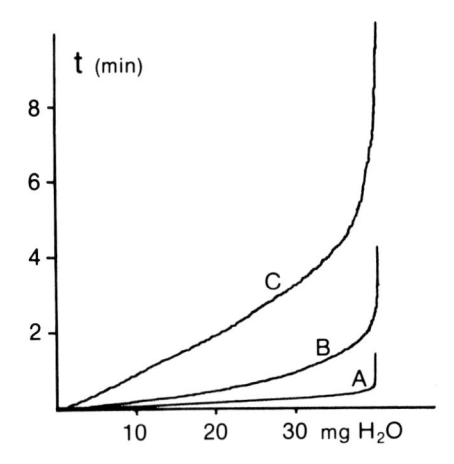

Fig. 3-1 Titration of 40 mg of water with different KF reagents [8102].
A: using a two-component reagent
B: using a pyridine-free one-component reagent
C: using a pyridine based one-component reagent

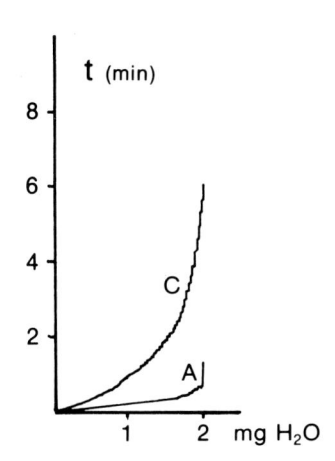

Fig. 3-2 Titration of 2 mg of water with different KF reagents [8102].
A, C as in Fig. 3-1

tion is in error, then the following water determination is likewise in error. Figures 3–1 and 3–2 clearly show that a two-component reagent is to be preferred.

The differences in end-point stability are also shown by both diagrams. The two-component reagent gives a stable end point, which is not influenced by the equipment used. A pyridine based one-component reagent on the other hand shows no signs of a real end point although the titrator has stopped automatically (curve C). The reagent consumption and the analytical results depend to a certain extent on the particular conditions being employed.

A further advantage of a two-component reagent is its shelf life. In a tightly closed bottle the titrant component is absolutely stable and the standardisation of the titre can be reduced to a periodic check. Likewise the solvent component has an unlimited shelf life provided the solution was properly made up.

The two-component reagent exhibits a higher reactivity, which can sometimes be a disadvantage, too. A few interfering side reactions like the methylation of aniline, which can take place during the water determination in aniline, also will be accelerated. A one-component reagent is to be preferred in these cases.

Another fact has to be taken into account when using two-component reagents. Due to the high reaction speed, the drop of titrant added is immediately consumed and may not reach the indicator electrodes at all or not early enough. The titrator may then accelerate the addition of titrant to such a degree that overtitration results. For this reason a good stirrer is needed to mix the working medium rapidly, and the delivery rate of the titrator has to be preset accordingly.

Using the two-component reagent, a few modifications in working technique are also possible. Instead of using a pure solvent component, mixtures with other solvents can be used as working medium. For example, when fats are being investigated, up to

70% of chloroform or propanol can be added. A complete replacement of the solvent component is of course not possible. Sufficient sulphur dioxide and amine must always be available for the reaction with the amount of water to be titrated. The "water capacity" of the solvent component has to be taken into account. Back-titrations as well as extraction methods are also possible with two-component reagents. Titrations at elevated or lower temperatures are also possible. The "successive titration" must be carried outh with circumspection in order to avoid a deficiency of sulphur dioxide, amine or even methanol in the working medium. Occasionally, further solvent component has to be added.

In the light of the advantages mentioned, one can expect that the two-component technique will be used more extensively, mainly because of the accuracy which can be achieved.

3.3 Coulometric Determination

The coulometric determination of water was first intensively investigated by Meyer and Boyd [5901]. They developed a simple apparatus, tested a suitable reagent and examined several possible fields of application. In the last few years, coulometry has emerged as a significant alternative to volumetric methods. Due to the now commercially available analytical equipment, it is very easy to use and can give reliable analytical results.

The Karl Fischer reaction also is used in the coulometric determination of water. Iodine, however, is not added in the form of a volumetric solution but it is produced in an iodide containing solution by anodic oxidation:

$$2\,I^- - 2\,e \rightarrow I_2 \tag{3.1}$$

The principle is relatively simple. The reaction cell (Fig. 3–3) consists of a large anode compartment A and a small cathode compartment C which are separated by a diaphragm D. Each of the compartments has a platinium electrode EA and EC which conduct current through the cell. The water determination is carried out in the anode compartment. The anolyte is stirred and the current switched on. Iodine is produced at the anode EA which then immediately reacts with the water present in the compart-

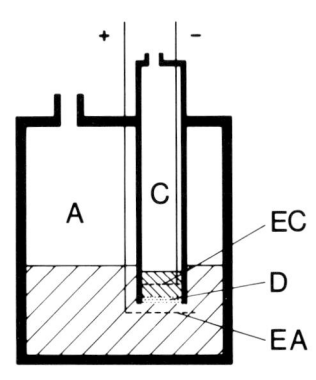

Fig 3-3 Coulometric titration cell. Depicted schematically according to [Hyd. p. 10]

ment. When all the water has been consumed an excess of iodine occurs, which can be detected voltametrically thus indicating the end point. The electric current is switched off automatically and the water content is calculated from the amount of current consumed.

To carry out a coulometric titration, the anode compartment is filled with the anolyte. This comprises of sulphur dioxide, a suitable amine, a soluble iodide and a solvent which is usually methanol (see Section 4.2). The cathode compartment is filled with the catholyte, which is a reagent of similar composition. Both solutions must be dehydrated prior to use. This requirement applies to the cathodic compartment in particular as otherwise any inherent water will slowly diffuse through the diaphragm into the anodic compartment causing delayed end-point adjustments and falsified results.

After the compartments have been filled with reagent, the current is switched on and the anodic compartment is dehydrated automatically. This dehydration must be carried out very carefully since the smallest amount of water introduces errors, which become very significant as coulometry is highly sensitive.

The sample is introduced into the dry anolyte by means of a syringe and the determination is started by "push button". It then runs automatically to the end point. The water content is indicated digitally in µg or on some instruments in mass units (percent or parts per million). Calibration of the instrument is not necessary as the current consumed can be measured absolutely.

In comparison to the volumetric Karl Fischer titration, coulometry is a micro-method. As the method utilises extremely small amounts of current, a maximum of 5–10 mg of water can be determined if the time for one analysis is restricted to approx. 10 minutes by economical reasons; it is predominantly used for substances with a very low water content (0.1–0.0001 %).

Because of its high sensitivity, atmospheric moisture must be absolutely excluded from the system. Coulometric cells are sealed units and must not be opened to introduce the sample. Samples are always introduced by means of injection with a syringe through a septum. This requires liquid samples. For the same reasons, the KF solution cannot be changed after each determination. Changing the solution, moisture penetrates into the system, which then has to be eliminated by pre-electrolysis. This can take several hours. It is therefore practical to fill the cells the evening before and dry them overnight. The individual determinations can then be successively carried out in the same reagent solution. A requirement for this method is, however, that each component of the "daily ration" is compatible with the other components and that no side reactions take place. This method is particularly suited to chemically inert substances like hydrocarbons, alcohols, ethers and esters.

The coulometric Karl Fischer titration is particularly to be recommended provided it meets the following 3 requirements:

1. The samples are liquid.
2. The samples have a low water content (0.0001–0.1 %).
3. The "daily ration" is as homogeneous as possible to reduce any detrimental effects.

Furthermore, there are a number of possibilities to investigate gaseous or solid samples. Gases can be introduced into the cell by means of a suitable gas inlet tube. For the determination of solids, the evaporation technique can be applied. The sample is heated in a tube and the water evaporated is carried into the cell by a stream of a dry gas or air. There it is absorbed in the anolyte and determined. In many cases, it is also possible to dissolve the solid in a suitable solvent. Besides alcohols, non-hygroscopic solvents like chloroform are preferred to dissolve solid samples. These solvents have a low solubility for water and minimize the errors caused by air humidity. From insoluble solids, moisture can be extracted using solvents. The extracts are then injected into the coulometric cell.

In comparison to volumetric titration, coulometric titration is more sensitive. It permits the determination of smaller amounts of water but it is equally sensitive to disturbances caused by chemical side reactions. The exclusion of atmospheric moisture is absolutely necessary. This applies not only to the coulometric cell but also to handling of the sample [5901]. The syringes must be absolutely dry as otherwise moisture will penetrate rapidly into the sample. For the determination of 10 ppm water in benzene, Swensen and Keyworth [6301] recommended to dry the sample bottles at 160 °C, to seal them with a septum while still warm and to cool in a nitrogen atmosphere. Coulometry appears to be more sensitive to chemical side reactions like esterification of acids, formation of acetals and ketals, or methylation.

Only a handfull of publications on the applications of coulometry are available to date. It is to be expected, however, that this analytical technique will considerably extend, as the handling involved is simple and the results obtained are reliable.

3.4 Back Titration

In volumetric analysis, back titration is made use of when a direct titration runs too slowly. As the Karl Fischer titration has a delayed end-point adjustment, the back-titration method is employed, as well. Using this technique, an excess of KF reagent is added to the sample or to the sample solution and after a short time, the excess is back-titrated with methanol normally containing about 5 mg of water per ml. This method can be used with both one- and two-component reagents.

The advantages of using a back titration method are seen in the better, quicker and more stable end-point adjustment. When titrating samples which are essentially free of water, direct titration requires perhaps only a few drops of reagent and the end point is delayed. By adding an excess of KF reagent and titrating back, at the end point a higher amount of sulphur dioxide is present in the system, which ensures a better end-point adjustment. The same effect can be achieved by adding water in the form of aqueous methanol to the working medium at the pre-titration step. This way also ensures a higher sulphur dioxide concentration at the end point and a better end-point adjustment. When titrating with a two-component reagent, an acceleration of the reaction is not necessary. The back titration is advantageous for pyridine-based one-component reagents only in certain exceptional circumstances.

A further cause of a delayed end point is penetrating moisture from outside. The end point keeps vanishing so that a delayed equilibrium adjustment might be assumed. Employing a back titration in this case, the end point would be stable as the interference caused by the penetrating moisture and the water added during the back titration would have the same effect. Such a "stabilized" end point does not increase the accuracy at all. It merely masks the source of the error. It would be more advisable to eliminate the source of the error which is more apparent in a direct titration.

The same considerations apply to determinations of insoluble substances. In this case the water must be extracted during titration. A vanishing end point indicates that the extraction is not yet complete. Using the back titration, the incomplete extraction is covered. Sometimes an excess of KF solution is applied to accelerate the extraction of water. Recent investigations of foods, undertaken by Zürcher and Hadorn [7801], indicate, however, that this influence is relatively insignificant.

Back titration had certainly definite advantages in the early days of KF titration when non-alcoholic organic liquids were titrated directly with the KF solution. The methanol content at the equivalence point presumably was too low so that a back titration was more or less inevitable. This mode is of purely historic interest now.

The back titration method also has its disadvantages. Firstly, it must be pointed out that more equipment, more chemicals and more time are necessary. For a back titration, two burettes and two reagent solutions are needed – the KF reagent and the water-in-methanol solution. A titre determination for each solution is also necessary. This doubles the amount of work involved and the risk of incorrect standardisations. The interferences due to atmospheric moisture are also greater as both solutions are hygroscopic.

Comparing advantages and disadvantages of the back-titration method for KF determinations, it can be concluded that the back titration should remain the exception and only be employed in well explained circumstances.

3.5 Titrations at Different Temperatures

The KF titration is normally carried out at room temperature and many procedures recommend to cool to 20 °C if the sample had been pre-treated at an elevated temperature. There is no inherent reason for carrying out titration only at 20 °C. Verhoef and Barendrecht [7703] found the reaction rate constant to exhibit a normal temperature dependance. It is reasonable to assume that titration can be carried out at both, higher and lower temperature, without any effect on the stoichiometry of the reaction. Thus, it is possible to titrate at different temperatures if the substance under investigation requires a modified treatment.

Substances which dissolve only slowly in methanol or other KF working media cause longer titration times. It takes at least one hour to dissolve sugar and this causes errors due to atmospheric moisture which have to be compensated for [7102]. A titration at an elevated temperature offers the possibility to avoid such problems. At 50 °C saccharose dissolves within 3 min and the titration does not take much longer [8302].

The extraction of moisture from insoluble samples at increased temperatures is particularly advantageous and this means of sample treatment is very often applied to foods (see Chap. 9). Here, a titration at 50–60 °C offers the possibility of combining the advantages of extraction at an increased temperature with those of a direct titration. Seibel and Bolling [5802, 5803] used such a procedure for their investigations of wheat, as did Zürcher and Hadorn [7801, 7804, 8008, 8106, 8207] with a variety of foodstuffs. We also tested its applicability to foods [8302].

Lower temperatures can be applied to eliminate unwanted side reactions. A well-known example is the determination of water in aldehydes and ketones (see Section 7.5). The formation of acetals and ketals can be inhibited to a certain degree by cooling to 0 °C or −10 °C thus enabling a direct titration of water.

The temperature cannot be varied unlimitedly, but the limits have not yet been established. Even at −30 °C, titrations can still be carried out at an acceptable rate of reaction. This is especially true of pyridine-free reagents which are more reactive. At increased temperature the KF reagent tends to decompose and produces higher blank values (see Section 4.3). Apart from that, an upper temperature limit is set by the boiling point of methanol, which means that 50 °C or 60 °C is the maximum working temperature for all practical purposes. When working at increased temperatures, it is advisable to check the blank value of the titration cell at the same temperature, as not only the solvent but also the KF reagent can change the rate of decomposition.

Titrating at different temperatures does not necessitate large-scale technical equipment. The titration cell can be cooled to the desired temperature in a few minutes by placing it in an ice bath or other suitable coolant. Heating can be accomplished by using an oil bath or suitable hot plate, which ideally should be equipped with a magnetic stirrer. If it is intended to work frequently at different temperatures, a titration cell equipped with a thermostatically controlled water jacket is the most reliable method of maintaining a constant temperature.

3.6 Continuous Determinations

To save a production process a permanent quality control can prove very advantageous and a few methods for continuous water determination will be described.

Kågevall, Åström and Cedergren [8001] developed a method of injecting samples into a continuous flow of KF solution (flow-injection analysis). The change in concentration of KF solution, specially developed for this purpose is monitored either spectrophotometrically or potentiometrically.

Koupparis and Malmstadt [8202] employ a stopped-flow analyzer and a two-component reagent. 2 ml of the solvent-component and 0.5 ml of the liquid sample are

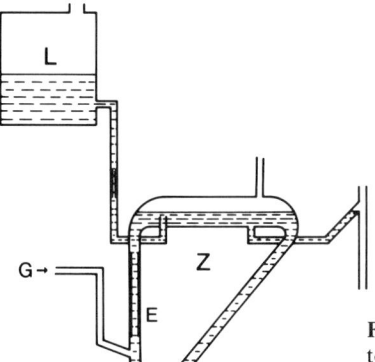

Fig. 3-4 Coulometric titration cell for continuous determination of water in gases. Depicted schematically according to Sistig [7103]

mixed and 0.15 ml of the titrant-component are added. After a period of 10 seconds the remaining concentration of iodine is measured photometrically.

For the continuous determination of water in gases the following coulometric measuring system, as depicted in Fig. 3–4, was recommended by Sistig [7103]. A stream of gas, G, is introduced into a closed-system cell arrangement Z. The water is absorbed by the KF solution. The ascending part of the cell contains a generator electrode E and an indicating system and here the water is determined coulometrically. The intensity of the current is a measure of the water content and can be recorded. Vessel L containing a modified KF solution and an overflow arrangement ensures a constant level in the cell. The complete system is absolutely closed to prevent the penetration of atmospheric humidity.

Certain commercially available coulometers are adaptable for use as continuous analyzers. The "drift" indicator of the Metrohm KF Coulometer 652 for example, could be used as a direct reading of the water content in a gaseous stream flow.

3.7 Indication

3.7.1 Visual Indication

The Karl Fischer solution is intensive brown in colour, and when titrating a solution with the KF reagent, a definite brown colouration becomes visible as soon as the equivalence point has been surpassed. A KF titration is thus self-indicating and permits visual indication.

By comparison to other types of titrations, however, this colour change can pose several problems. As the colour of the working medium turns yellow during titration and the intensity of this yellow background colour increases as the end point is approached, the actual colour change is really from yellow to brown which can limit the evaluation.

According to Verhoef and Barendrecht [7703, 7704] this yellowing is caused by the chromophore SO_2I^- which is thus formed:

$$SO_2 + I^- \rightleftharpoons SO_2I^-. \tag{3.2}$$

The intensity of the yellow colouration depends on the iodine concentration and hence on the amount of water titrated. It also depends on the concentration of free sulphur dioxide and, therefore, on the pH of the working medium and possibly on the substance to be investigated. Shifting the equilibrium in Eq. (2.11) to the right the intensity of the yellow colouration decreases. The practical implications of this are discussed in Section 6.5.

A further effect of the KF titration must be considered. It is the tendency of the end point to vanish. Two factors are responsible for this reversal. The first is due to the time taken for a KF reaction to attain equilibrium in Eq. (2.22). The reaction rate constant according to Eq. (2.14) is lesser in acid medium, and equilibrium attainment takes longer. In a neutral solution the reaction proceeds more rapidly (see Sections 2.2 and 6.2). In practice, this equilibrium delay is counterbalanced by defining the end point. The end point is defined as one which remains stable for 5–20 s. The second reason for a vanishing end-point is the instrusion of atmospheric moisture which is continously absorbed by the titration medium (see Section 6.4). It is sometimes not easy to distinguish between an incomplete equilibrium and a constant infiltration of moisture.

3.7.2 Instrumental Indication

As visual indication is not free from problems, other possibilities to detect the end point were investigated. Photometric, conductometric and dilatometric indicating systems [7905] as well as potentiometric indications using bimetallic combinations were employed.

Indication using polarized electrodes is the preferred method which is employed generally. This is often referred to in the literature as "dead-stop indication" although this term is not strictly true. "Dead-stop indication" was first tested and applied to KF titrations by Wernimont and Hopkinson [4301]. The indicating system consists of two platinum electrodes, which are placed in the solution to be titrated. A constant voltage of 10–500 mV is applied between the electrodes. When a redox pair, in this case I/I^-, is present in the solution, a current of 30 µA for example, flows between the electrodes. If no free iodine is present in the solution, the cathodic reaction (reduction of iodine to iodide) is not possible – the cathode is "polarized" and the electric current is stopped. Thus, the ammeter suddenly drops to zero, i.e. "dead stop". This sequence only applies to the KF back-titration method as employed by Wernimont and Hopkinson.

A direct KF titration utilizes this sequence in reverse. The solution contains no free iodine before the equivalence point has been reached and no current flows through the solution. Surpassing the end point the current suddenly increases. The term "dead stop" is incorrect for this. Kolthoff used the term "kick-off" indication.

According to IUPAC this method of indication is termed "amperometry with two indicator electrodes", regardless of the direction of operational sequence. The term "biamperometric" is more usual in practice and will be used in this book for the sake of brevity.

The electrodes can also be polarized using a constant current, for example 1 µA. In this case the voltage between the two electrodes is measured and serves as the indicator. When polarized, voltages of 200–500 mV can be measured. When the equivalence point is reached, the voltage drops to virtually zero thus indicating the titration end point. This type of end-point indication is referred to by IUPAC as "controlled current potentiometry with two electrodes". This is shortened to "bipotentiometric" or the lesser used "voltametric" in practice.

Instrumental indication substantiates the realisation of the end point. The problems of a delayed adjustment of equilibrium must, however, also be taken into account. The practical aspects of instrumental indication are treated in Sections 5.1.1 and 6.5.

4 Reagent Solutions

The generally used term "Karl Fischer solution" creates the impression that there is only one type of solution given by Karl Fischer himself and applied throughout. But this is not true. Many different reagents have been reported in the last 50 years. Concentrations and mole ratios have been changed, solvents have been substituted and last not least the unpleasant pyridine has been replaced by other bases. The one-component reagents have been supplemented by two-component reagents and by coulometric reagents.

4.1 Reagent Solutions for Volumetric Titrations

In the opinion of Karl Fischer a mole ratio of $I_2:SO_2:C_5H_5N$ of 1:1:4 was necessary to complete the reaction of Eq. (2.2). However, in order to accelerate the reaction, he increased the amounts of sulphur dioxide and pyridine and recommended a methanolic solution comprising of:

Iodine (I_2)	0.20 mol/l
Sulphur dioxide	0.60 mol/l
Pyridine	2.0 mol/l

The mole ratio is thus 1:3:10. Fischer had calculated a water equivalent (WE) of 7.2 mg/ml but he found a considerably lower titre [3501]. Today we know this solution composition to have a WE of approximately 3.6 mg H_2O/ml.

Smith, Bryant and Mitchell proposed higher concentrations [3901]:

Iodine (I_2)	0.33 mol/l
Sulphur dioxide	1.0 mol/l
Pyridine	3.3 mol/l

This solution has a theoretical WE of 6 mg H_2O/ml, which is always somewhat reduced by moisture. This formulation is still used today, slightly modified, and it is recommended in "Reagent Chemicals" of the American Chemical Society for example. One disadvantage of this solution is its limited shelf life. After making up this solution, the titre decreases relatively rapidly, 10–20% in the first day and then successively at an approximate rate of 1% per day when stored in a closed container at room temperature. In spite of this, several similar formulations were proposed and tried out, nearly all of them are applicable in a similar manner [Aqu. p. 109–112].

In order to prolong the shelf life, it was proposed to make up the solution in steps. A large amount of a sulphur dioxide/pyridine/methanol solution is prepared first and iodine is added to a portion of this as required. This naturally led to the introduction of a two-component reagent by the manufacturers of laboratory chemicals. Component A is a solution of sulphur dioxide and pyridine in methanol, component B is a solution of iodine in methanol. Each component is stable in storage. They are mixed together just before use. Thus it was possible to produce and market the Karl Fischer reagent commercially. Today such solutions are supplied by every manufacturer and supplier of laboratory chemicals. The solutions as offered by different manufacturers, differ only very slightly in their chemical composition. Riedel-de Haën for example offers the solutions in the following compositions:

Solution A: Sulphur dioxide 3.2 mol/l
             Pyridine   4.5 mol/l
Solution B: Iodine (I_2)   0.6 mol/l

Mixing both solutions in equal volumes the user gets a Karl Fischer reagent with a theoretical WE of 5.4 mg H_2O/ml. Due to moisture and the inevitable loss of titre, the WE of the combined solution drops to approximately 4 mg/ml after 2 days. The supplier usually specifies a minimum WE of 3 mg/ml.

In 1947, Johansson recommended using both solutions separately [4701] and this led to the introduction of the two-component technique (see Section 3.2). He also proposed different concentrations:

Solution A: Sulphur dioxide 1.5 mol/l
             Pyridine   6   mol/l
Solution B: Iodine (I_2)   0.11 mol/l

Solution B corresponds to a WE of approximately 2 mg/ml. Other strengths can be made up as desired. In spite of its obvious advantages (see Section 3.2), the two-component technique has remained unduly ignored for too long.

Significantly more recognized was the proposal by Peters and Jungnickel [5501] to replace the methanol in the one-component reagent by 2-methoxyethanol. The shelf life of a reagent made up with 2-methoxyethanol is much longer and it is less sensitive to side reactions, like those with ketones for example. Peters and Jungnickel proposed the following compositions:

Iodine (I_2)   0.5 mol/l
Sulphur dioxide   1.5 mol/l
Pyridine   5   mol/l

Thus, the mole ratio of 1:3:10, as originally proposed by Fischer, was retained. The titre decreases at a monthly rate of approximately 2% according to present-day knowledge. This reagent can be stored and used up to 2 years and it is offered, slightly modified, by every manufacturer and supplier of laboratory chemicals, now; sometimes specified as "stabilized". Several compositions available on the market are summarized in Table 4–1.

Although the chemical composition differs from producer to producer, the technical application is hardly affected. Increasing the pyridine content can accelerate the KF reaction to a certain degree. A commercially available reagent specified as "rapid" has the following composition:

4.1 Reagent Solutions for Volumetric Titrations

Iodine (I$_2$)   0.35 mol/l
Sulphur dioxide   1.2 mol/l
Pyridine   8.4 mol/l

Table 4-1. Commercially available one-component reagents (analysed by the author)

	Merck (1979)	Baker (1979)	Riedel-de Haën (1982)	Mallinckrodt (1975)
Iodine (I$_2$)	0.35	0.2	0.35	0.35 mol/l
Sulphur dioxide	1.0	1.2	1.5	1.1 mol/l
Pyridine	3.3	4.8	3.2	2.6 mol/l

Reagents which essentially consist of pyridine and contain only a minimum of methanol are listed in Table 4–2. These reagents prove particularly advantageous for the investigation of ketones as the formation of ketals is inhibited by working in an alcohol-free medium. For aldehydes, however, the water content found is too low [Aqu. p. 383].

Table 4-2. Reagents low in methanol

	Wernimont and Hopkinson [4301]	Mitchell and Smith [Aqu. p. 383]
Pyridine	1700 ml	920 ml
Methanol	200 ml	30 ml
Sulphur dioxide	380 g	64 g
Iodine	500 g	84.7 g

Dimethylformamide is also used as a solvent for KF reagents. Klimova, Sherman and L'vov [6703] recommended the following reagent composition for carbonyl compounds and those compounds having no active hydrogen atoms:

Iodine (I$_2$)   0.2 mol/l
Sulphur dioxide   0.6 mol/l
Pyridine   1.3 mol/l

In these non-alcoholic solutions, the stoichiometry appears to deviate from the KF equation (2.22), and to tend towards the Bunsen reaction (2.2). Sherman [8006] found the titre of the solutions to depend on the solvent which is used as the working medium.

Mitsubishi applied for patents using alkyl carbonates as the solvent [8005]. Examples of these are (for 1 l of reagent):

Iodine (I$_2$)	0.16 mol	0.33 mol
Sulphur dioxide	0.5 mol	1.0 mol
Pyridine	1.7 mol	3.4 mol
Ethylene carbonate	418 ml	617 ml
Chloroform	417 ml	–

One notable advantage of these solutions is the stability which is referenced as 10% titre loss in 300 days.

Pyridine, which is used as base in all classic KF reagents, has a very unpleasant intrinsic odour, which does not promote the use of KF reagents. It therefore seems surprising that only a few investigations have been undertaken to replace pyridine by other amines. Smith, Bryant and Mitchell [3901] reported that quinoline, aniline, dimethylaniline, tributylamine and triethanolamine could be used "although not with equal success". An aniline-based reagent decomposes within 24 hours. Dimethylaniline proved somewhat better. A reagent made up with triethanolamine decomposed during preparation. Pyridine was thus described as "most satisfactory" and was continued to be used. Johansson [5601] carried out similar investigations with his two-component compositions. Aniline and ethanolamine were tried but were found to give no stable end point. The only base which gave reproducible end points was found to be hexamethylenetetramine. Its application, however, proved unsatisfactory on account of its insolubility and the poor accuracy.

Van der Meulen used "alkali alcoholates, phenolates, acetates" instead of pyridine to neutralize the acids liberated during the KF reaction and patented this reagent [5301]. One example is a methanolic solution of the following components:

Iodine (I_2)   0.5 mol/l
Sulphur dioxide   1   mol/l
Sodium acetate   2   mol/l

Sherman, Zabokritski and Klimova [7302] pointed out that precipitation could occur with this solution and they recommended lower concentrations:

Iodine (I_2)   0.25 mol/l
Sulphur dioxide   0.5 mol/l
Sodium acetate   1   mol/l
Potassium iodide 0.3 mol/l

A two-component reagent named "Reaquant" was introduced by J.T. Baker in 1979 and was also based on sodium acetate. It had the following composition:

Solvent-component: Sulphur dioxide 0.5 mol/l
                     Sodium acetate   1   mol/l
Titrant-component: Iodine (I_2)   0.2 mol/l

A patent for a variation of this, in which methanol is partially replaced by 2-methoxyethanol, was applied for [7904]. The pH is adjusted to approximately 6 by sodium acetate and a nearly colourless solvent component is obtained which is an advantage for visual end-point indication. The higher reaction rate is an added advantage as this ensures a rapid titration. The shelf life of the reagent is unsatisfactory, however. Acetic acid is liberated by the methyl sulphurous acid and esterifies with the methanol. The water produced by the esterification decomposes the reagent slowly and it can be stored only for a limited time at room temperature. Also, solids can precipitate during storage. The "Reaquant" reagent was later modified by using sodium salicylate and other bases.

We found aliphatic amines to be suitable as bases, provided a mole ratio of amine to sulphur dioxide is chosen to maintain and to stabilize the pH-value of 5–8. Two-component reagents as well as one-component reagents can be produced using these amines [8002, 8103]. Riedel-de Haën introduced such a reagent to the market in 1980. The two-component reagent – Hydranal-Solvent and Hydranal-Titrant – is distinguished by a high reaction rate, a stable end point and accurate results [8102]. The titrant component is offered with a guaranteed WE of 5.00 mg/ml. Both components have an unlimited shelf life. The composition of the reagent is protected by patents [8003].

Pyridine can also be replaced by other heterocyclic bases. Imidazole for example, has a higher basicity than pyridine and is much more suitable to maintain the pH of the KF system in the optimum range (see Section 2.2). Both two-component reagents as well as one-component reagents can be made using heterocyclic bases. Patents for these bases have been applied for by Riedel-de Haën and a reagent is on the market under the trade-name Hydranal-Composite. Due to the optimum pH range, Hydranal-Composite has a higher rate of reaction and gives more accurate results than the conventional one-component reagents [8102]. Similarly, its stability during storage is better than that of other reagents containing 2-methoxyethanal as solvent.

These recently introduced bases have altered the character of the Karl Fischer reagent. The slowly proceeding titration, a sluggish attainment of equilibrium following each addition of reagent, and the uncertainty whether the end point has actually been reached, are no longer apparent. The new bases act more alkaline than pyridine, the sulphur dioxide in Eq. (2.11) is fully transformed to methyl sulphite and the reaction in Eq. (2.22) is thus accelerated accordingly. A comparison of the titration times using commercially available reagents is shown in Figs. 3–1 and 3–2. The Bunsen reaction in the equilibrium of Eq. (2.21) is presumably suppressed to such an extent that the stoichiometry follows the KF equation (2.22). Table 4–3 shows that a much greater accuracy can thereby be achieved.

Table 4-3. Errors and scatter of water determinations according to [8102]

Water	Conventional KF reagent	Hydranal-Composite	Hydranal-Solvent Hydranal-Titrant
100 mg	± 0.44 %	± 0.30 %	± 0.20 %
50 mg	+ 0.56 ± 0.62 %	+ 0.06 ± 0.24 %	+ 0.01 ± 0.30 %
20 mg	+ 1.68 ± 0.93 %	+ 0.33 ± 0.60 %	+ 0.02 ± 0.25 %
10 mg	+ 2.59 ± 1.85 %	+ 0.84 ± 0.6 %	− 0.1 ± 0.5 %
5 mg	+ 2.40 ± 2.5 %	+ 0.72 ± 1.2 %	+ 0.2 ± 1.0 %

Thus, the new reagents, which were only designed to replace pyridine, are greatly superior to their classic standard. The absence of the unpleasant smell of pyridine turned out to be only an additional but welcome accompaniment.

Further newly published developments [8111, 8115] deal with the application of non-hygroscopic solvents to produce titrant components of two-component reagents. These titrants are not affected by atmospheric humidity thus ensuring a high degree of titre stability.

4.2 Reagent Solutions for Coulometry

For the coulometric determination of water, reagent solutions similar to the volumetric ones were initially used. A few investigations deal with the development of new formulations, mainly for the anolyte. The catholyte seems to act only as an auxiliary reagent and it is mostly underestimated. We did not find any investigation on the mechanism of the cathodic reaction.

A reagent differing from the standard KF reagent was patented by Oehme [5801]. It contains an alkyl iodide which is anodically oxidized to iodine.

Fundamental investigations on the coulometric determination of water were carried out by Meyer and Boyd [5901]. They used a "spent" KF solution as the reagent, i.e. a KF solution in which the iodine has already been reduced to iodide by the addition of water. They recommended the following composition:

Iodine (I_2)	0.28 mol/l
Sulphur dioxide	1.12 mol/l
Pyridine	4 mol/l

The mole ratio is increased to 1:4:14 so that the "spent" solution contains 0.84 mol/l of sulphur dioxide and 3.26 mol/l of pyridine corresponding to the "classic" mole ratio of 1:3:10. Both cathodic and anodic compartments are charged with the same solution. The electrolyte in the cathodic compartment was found to be not completely satisfactory as reducing substances are produced, which diffuse through the diaphragm into the anode compartment and react with the iodine. This leads to higher water contents being indicated.

Swenson and Keyworth [6301] used formamide instead of methanol as a solvent. They stated a higher rate of reactivity of the reagent and its suitability for water determination in ketones. The same reagent is used by Bizot [6702]. He disclosed the following composition as an anolyte:

Potassium iodide (KI)	0.027 mol/l
Sulphur dioxide	0.5 mol/l
Pyridine	approx. 9 mol/l
Formamide	approx. 200 ml

After the addition of 30–150 mg/ml of azobenzene, the same solution is used for the cathode compartment thus avoiding the generation of reducing substances. The reagent was applied for water determination in acetic acid (18 ppm H_2O), benzene (32 ppm), dimethylformamide (14 ppm) and many other liquids. Different cells for gases and solids are also described.

Surprisingly, an alcohol-free system is used here which can not form an alkyl sulphite. Swenson and Keyworth formulated a reaction sequence based on Eqs. (2.5) (2.6) by Mitchell and Smith, in which formamide takes the place of methanol:

$$PySO_3 + H_2NCOH \rightarrow Py \cdot HSO_3HNCOH \tag{4.1}$$

Miyake and Suto [7408] recommend a solution of potassium iodide and sulphur dioxide in pyridine as the anolyte, diluted with methanol. Either potassium iodide in methanol or sulphur dioxide in methanol/pyridine is used as the catholyte. To analyse ketones, they recommend an anolyte containing acetonitrile as a solvent [7705]. Cedergren [7401] uses a methanolic solution consisting of:

Iodine (I_2)	0.1 mol/l
Sulphur dioxide	0.6 mol/l
Pyridine	1.0 mol/l

AURORA UNIVERSITY
CIRCULATION DESK
630-844-7583

This book has been obtained for your use through the courtesy of another library. As the borrower of this material, you are subject to the loan and fine policies of the library that has loaned this material.

If you wish to renew this item, please contact the library at least 3 days before the due date, or you may renew the item online following the directions below.

Renewing an item from the OPAC

1. Access the library's home page www.aurora.edu/library
2. Open "Illinet Online."
3. Open "Aurora University Library Catalog."
4. Open "My Account"
5. Select your home library (Aurora University)
6. Enter your library ID (2251100+your University ID number)
7. Enter your last name
8. Click on Log-in
9. When your record comes up, click in the renew box(es) to the left of the item(s) you wish to renew. If your record does not come up, contact the library at 630-844-5439.
10. Click on "renew selected items"

Patron ___Holland, Cara___

Due Date ___07/27/2010___

4.2 Reagent Solutions for Coulometry

To use this solution as the anolyte, the iodine is reduced to iodide by the addition of water. The catholyte is the same solution in its original composition. The reduction of iodine to iodide serves as the cathodic reaction.

Commercially available reagents are in approximate agreement with this composition. Merck supplies the following reagent (analysed in 1982):

Iodine (I_2)   0.03 mol/l
Sulphur dioxide   0.5 mol/l
Pyridine   1.8 mol/l

The solvent is methanol. The same solution is used both as anolyte and catholyte.

Mitsubishi supplies for its "Moisture Meter CA 02" two different electrolytes for the anode (A) and the cathode (C) compartment. They are named Aquamicron A and Aquamicron C. We found the following components, dissolved in methanol (analysed in 1982):

	Aquamicron A	Aquamicron C
Iodine (as I_2)	0.1 mol/l	–
Sulphur dioxide	0.5 mol/l	1.0 mol/l
Pyridine	2.5 mol/l	2.0 mol/l
Chloroform	260 g/l	–
Carbon tetrachloride	–	600 g/l
Propylene glycol	170 g/l	–

For its KF coulometer "Aquatest IV", the Photovolt Corporation (USA) supplies three solutions named A, B and "Generator Solution". Their preparation is mentioned in ASTM D 1533–79 [7908]. They have the following composition:

	Solution A	Solution B
Iodine (I_2)	–	0.38 mol/l
Sulphur dioxide	2.7 mol/l	0.7 mol/l
Pyridine	10 mol/l	0.6 mol/l
Methyl glycol	–	370 g/l

Methanol serves as solvent. The anolyte is made by mixing 75 ml of solution A with 105 ml of solution B and decolourizing this by addition of water. The "generator solution" serves as the catholyte and is identical to solution B.

For coulometry Riedel-de Haën has developed pyridine-free reagents. The patented bases [8003, 8010] give reagents for rapid determinations, stable end points and with a high degree of accuracy. Hydranal-Coulomat A is the anolyte and contains both methanol and chloroform as solvents and has a high solubility for non-polar substances. Hydranal-Coulomat C, the catholyte, is a solution of similar composition which prevents the formation of reducing substances.

4.3 Stability of Solutions

"If the iodine solution is kept in a well-sealed bottle, the titre slowly decreases in a few weeks, probably due to the substitution of hydrogen by iodine." Fifty years ago Karl Fischer [3501] described the behaviour of his reagent solution on storage. This observation is still valid and our present-day knowledge is only a little more than the information conveyed in the old statement from Fischer. A decreasing titre of reagent solutions is often mentioned in the literature, but even giving precise decay rates can cause problems. The question of the mechanism of decay can not be answered.

A freshly prepared methanolic solution has only 80% of its theoretical value. The titre then drops further to 50% after a month and to 40% after 3 months [Aqu. p. 92]. The values are not reproducible. Eberius [Eb. p. 33] compared seven KF reagents from different manufactures (Fig. 4-1, curves A-F). He found very differing loss rates. The curves A to E show decay rates between 0.04 and 0.07 mg H_2O per day (as means over a period of 20 days). "Noticeable is the titre stability of solution F, prepared by just happening to use a pyridine of unknown but effective stabilizing constituents" [Eb. p. 34]. We found a titre loss of 35% in 9 weeks for a methanolic one-component reagent using diethanolamine as a base [8002].

Decay rates given in the literature are often to be treated sceptically, especially those from long-term stability investigations. The atmospheric moisture represents a considerable source of errors and it is not always excluded. For example, there are reports on the loss of efficiency in two-component reagents, although Seaman, McComas and Allen had stated in 1949 that these solutions are absolutely stable and a loss of titre is solely caused by the effect of moisture [4902]. The stability of these solutions, however, is common knowledge today.

The introduction of 2-methoxyethanol into one-component reagents by Peters and Jungnickel [5501] resulted in a significant improvement of the stability. The authors quote a rate of titre loss of 9% per month for this new reagent, which compares to a rate of 22% for a similar methanolic solution. Further details concerning stability are sparse. We found only one reference where a solution remained "essentially constant" for 20 days [Aqu. p. 94].

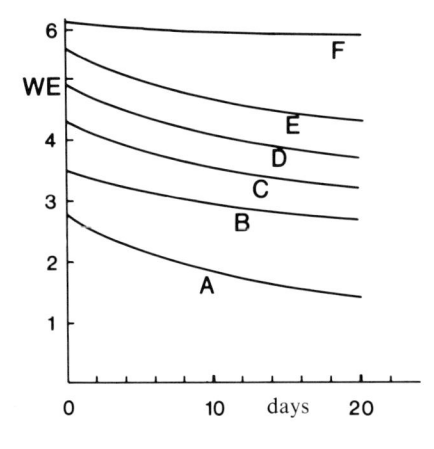

Fig. 4-1 Decrease of titre in different KF reagents according to Eberius [Eb. p. 33]

4.3 Stability of Solutions

Table 4-4. Change of titre of the one-component reagent, Riedel-de Haën product No. 36115 (Pyridine, sulphur dioxide, iodine in 2-methoxyethanol)

3 months	−11%
6 months	−19%
9 months	−22%
12 months	−24%
15 months	−32%
18 months	−36%
24 months	−44%

We undertook several studies on stability of our own. A one-component reagent using 2-methoxyethanol as the solvent stored in sealed glass ampoules exhibited a titre loss over a period of two years as shown in Table 4-4.

Table 4-4 gives an average value of −1.9% per month. Using diethanolamine as base, we found a rate of approximately −2% [8002]. In further studies over periods of 6 months, we calculated a monthly rate of −2.1% for pyridine, −3.1% for morpholine, −1.2% for diethanolamine, −0.4% for imidazole and −0.9% for 2-methylimidazole.

A reagent containing pyridine as base and propylene carbonate as the solvent was quoted in a patent application by Mitsubishi of having a decay rate of about 1% per month [8005].

Investigations to elucidate a reaction mechanism for titre decay have apparently not been published in the last years and earlier studies appear to be very uncertain. Eberius [Eb. p. 44] pointed out that boiling pure pyridine with iodine produces a dark-brown substance, presumably an iodopyridine as a result of nuclear substitution. "If such a consumption of iodine ... occurs slowly in the KF solution..." this could be a possible explanation [Eb. p. 44]. An iodofication, however, only occurs in pure pyridine; nobody has detected iodopyridine in decomposed KF solutions. Additionally, KF solutions containing diethanolamine, morpholine or imidazole instead of pyridine, show a similar loss of titre. Is it possible that these amines are also iodofied in a similar manner? This hypothetical iodofication appears not to suffice as a general explanation.

Smith, Bryant and Mitchell [3901] distilled a three-month old KF solution and found only 93% methanol and 90% pyridine originally used in the preparation of the solution. After distillation from excess sodium hydroxide "a small amount of distillate with a very sharp amine odor somewhat like piperidine was obtained." As the same smell was noticed during the distillation of N-alkylpyridinium salts, an N-alkylation was presumed. The experimental data appear rather inadequate to us to derive a reaction sequence for this.

We ourselves [8002] found N-methylaniline and N,N-dimethylaniline in a decomposed KF reagent which had been made up with aniline as base. But aniline is not a suitable KF base so that these findings may not be relevant.

The hypotheses about the decomposition of KF reagents are not very convincing. They all assume a reaction of the pyridine. As pyridine-free reagents also show a similar loss of titre, they appear to be only partially applicable as an explanation for loss of titre. The same applies to hypothetical reactions of free sulphur dioxide.

Table 4-5. Titre change of KF reagents with imidazole and various alcohols when stored over a period of 3 months

Solvent	Loss of titre per month
Methanol	− 15 %
Ethanol	− 7.6 %
1-Propanol	− 6.3 %
1-Butanol	− 6 %
2-Butanol	− 5 %
2-Methoxyethanol	− 2.1 %
2-Ethoxyethanol	− 1.4 %
2-Butoxyethanol	− 1.9 %

Table 4-6. Titre change of KF reagents as a function of the mole ratio SO_2: amine (diethanolamine, iodine, sulphur dioxide, 2-methoxyethanol)

Concentration (in mol/l)			Titre mg H_2O/ml	Change after 3 weeks	
Diethanolamine	SO_2	I		at 20° C %	at 50° C %
2	1	0.8	4.47	− 86	decomposed in 2 days
2	1.5	0.8	5.08	− 28	− 70
2	2	0.8	6.10	− 6	− 33
2	2.5	0.8	6.20	− 1	− 30
2	3	0.8	5.89	− 20	− 75

Due to insufficient knowledge, a reaction for the decomposition of KF solutions can not be formulated. We however know a few factors which influence the rate of decomposition. The alcohol components shall be mentioned first. Loss rates of around 20% per month already have been noted for methanolic reagents. If 2-methoxyethanol is used, the rates are approximately 2% per month. We investigated a reagent containing imidazole as base and various alcohols as solvents and found the titre changes shown in Table 4–5. Although the results are not reproducible, the tendency is apparent: methanol shows a very high decay rate, and the stability improves with increasing chain length.

The rate of decomposition appears to be influenced by the pH or the mole ratio of sulphur dioxide to amine. For diethanolamine we found the values shown in Table 4–6. The best stability was found using a mole ratio of 1.25:1. Both acidic as well as basic solutions decomposed more rapidly.

Blomgren and Jenner [5702] improved the stability of reagents containing pyridine by the addition of a "stabilizing base", which should be stronger than pyridine.

Eberius' observation [Eb. p. 43] that a pyridine-free solution of sulphur dioxide and iodine in methanol decomposes very quickly can also be interpreted from the point of view that this solution is too acidic.

We don't know whether the bases used have an influence on the stability extending beyond that of the pH effect. We know there are bases not suitable, like aniline for ex-

4.3 Stability of Solutions

ample, which becomes alkylated in a KF reagent. We also know that pyridine, diethanolamine or imidazole show only very slight differences in their decay rates, whereby the pH value can not be completely ignored. One can assume that "suitable" bases have no direct influence, they appear to act only as buffers.

We did not find any investigation on whether the concentration of sulphur dioxide has an effect on the rate of decomposition.

If iodofication of pyridine causes the loss of titre, it would be expected that the decay rate is proportional to the concentration of iodine. But we have not found any publication concerning the influence of iodine on the decomposition. Some results have been brought in connection with other investigations. Blomgren and Jenner [5702] have carried out tests on the stability of KF reagents, from which we can conclude indirectly that the iodine concentration does not influence the decay rate. The same results were found by Cedergren [7402] investigating the kinetics of the coulometric water determination. At the end point he found a "rate of formation of water" of 6 μg H_2O per minute in 100 ml of reagent solution. This corresponds to a water formation of 86 mg per day for a 1 l KF solution. This would infer a decomposition rate of 1.7% per day for a reagent with a WE of 5 mg/ml. This rather surprising agreement fits in with the assumption that the formation of water in coulometric reagents and the titre loss of a KF reagent are caused by the same reaction.

Cedergren found that the formation of water in a coulometric reagent is not influenced by the iodine concentration and only slightly by the pyridine concentration. And perhaps here only the higher pH caused by the higher pyridine concentration is the reason. According to Cedergren, the quality of the pyridine is of much greater significance, which however is not further explained.

On the other hand, the presence of iodine is an absolute prerequisite for the decay reaction to take place, as the solvent components which contain sulphur dioxide, base and methanol are absolutely stable during storage.

The observation of Blomgren and Jenner [5201] that an increasing content of iodide reduces the decomposition of KF reagents, should also be mentioned. Possibly iodide is oxidized to iodine by atmospheric oxygen, compensating thus the proceeding decomposition. A mole ratio of $I^-:I_2$ of 3.2 is the optimum. Higher contents of iodide increase the titre.

Eberius [Eb. p. 39] found that an excess of methanol stabilizes the KF reagent. He based this on the supposition of a delayed formation of iodopyridine. But this stabilization could also be caused by the increasing iodide content produced by the water present in the methanol.

There is a rapid decay of KF reagents under alkaline conditions. We found [8002] a KF reagent containing isopropylamine as the base and having a pH of 9.2 to decompose during preparation. A reagent containing morpholine (pH 7.7) decomposed within one week. The decomposition is not caused using unsuitable amines. Our studies with diethanolamine [8103], reproduced in Table 4–6, indicate a rapid decomposition entering the alkaline range. These results also explain the unstable solutions obtained by Smith, Bryant and Mitchell [3901] replacing the pyridine by other amines.

Additional information on alkaline decomposition was provided by titration with a two-component reagent [8002]. Using a solvent component consisting of isopropylamine, with a pH of 9.2, for the titration of water results in a too high consumption of iodine solution and an unstable end point. It could be assumed that a part of the iodine is consumed for a side reaction. The side reaction proceeds very rapidly and it is possibly the disproportionation of iodine.

When attempting to compile all the individual results, a complex picture is obtained. More than one mechanism seem to be responsible for the decomposition of a KF solution:

1. The KF reagent decomposes when an unsuitable base is used, aniline for example. Pyridine might be classified as a suitable base, as a reaction of pyridine with iodine never has been detected. Possibly it is "unsuitable" if it contains detrimental substances like picoline and lutidine.
2. A second decay reaction appears to take place in the acidic medium between sulphur dioxide, iodine and alcohol. The rate of this reaction depends on the pH: the rate decreases with increasing pH. The base has no specific influence on this and neither has the iodine concentration, apparently, although iodine must be present. Possibly the alkyl sulphurous acid takes part in this reaction. A slow esterification to an alkyl sulphite followed by the removal of the water by a KF reaction could be one explanation.
3. It is unlikely that the acidic decomposition reaction is reactivated by attaining or going beyond pH 8, so that a different mechanism for alkaline decomposition must be assumed.

These three reactions for the decomposition are purely hypothetical and are hardly substantiated by results of actual studies. They would appear, however, to represent an improvement on the otherwise generalized considerations of "decomposition". If they stimulate interest to investigate further, then they have fulfilled their purpose even if contradictory results are thereby obtained.

4.4 Preparation of Solutions

Karl Fischer reagents are nowadays prepared commercially on a large scale, and despite their high costs, are cheaper than making the solutions oneself, as this requires a lot of experience. Furthermore, it should be pointed out that many formulations for pyridine-free reagents have been patented and self-made preparations would constitute an infringement of patent rights. The methods of preparations are mainly of historic interest now and, thus, only three of the well known methods are included here.

Preparation according to Fischer [3501]: 254 g iodine are dissolved in 5 l methanol. 790 g pyridine and 192 g liquid sulphur dioxide are then added. The solution has a WE of approximately 3 mg H_2O/ml.

Preparation according to Mitchell and Smith [Aqu. p. 109]: In a 9 l pyrex flask, 762 g iodine (3 mol) are dissolved in 2.4 l pyridine (29.7 mol) and 6 l methanol are added. This solution can be stored indefinitely. When required, 3 l of this are filled into the supply vessel of the burette and cooled with ice. 135 ml sulphur dioxide are condensed in a dry-ice/methanol cold trap and kept cold. The liquid sulphur dioxide is added to the methanol solution in the supply vessel and shaken. After cooling to room temperature, the solution is ready for use. The solution has a WE of approximately 5 mg H_2O/ml.

4.4 Preparation of Solutions

Preparation according to Eberius [Eb. p. 28]: 84 g iodine are dissolved in 530 ml methanol. 64 g liquid sulphur dioxide are added to 265 g pyridine whilst cooling with ice. This solution is then added to the methanolic solution. The solution should be syphoned as "shaking, splashing and handling using wide-necked vessels reduces the efficiency by up to 50%..." of the potential WE of approximately 6 mg H_2O/ml (Eberius).

The sequence in which the individual components are added differs from procedure to procedure. We prefer to mix methanol and pyridine first and to add sulphur dioxide whilst cooling. Iodine is added last. The sequence in which the four components are added together is completely arbitrary and has no influence on the quality or strength of the solution. Only a solution of sulphur dioxide and iodine in methanol in the absence of pyridine is unstable.

The choice of raw materials is normaly not critical. Methanol and 2-methoxyethanol are available in adequate purity. Their water contents do not exceed 0.1%. If a drying seems appropriate, molecular sieves should be preferentially used. Older procedures recommend drying the methanol with sodium sulphate or with copper sulphate and subsequent distillation. An original method of drying the methanol is to add sulphur dioxide and bromine, a variation of the KF reaction [Eb. p. 26].

Pyridine is also available commercially in adequate purity and with a sufficiently low water content (max. 0.1%). The homologues of pyridine, picoline and lutidine are undesirable though. They are possibly the cause of a rapid titre loss. Pyridine can be dried by adding benzene followed by azeotropic distillation. The azeotropic mixture water/benzene/pyridine boils at 67 °C, followed by benzene and then pyridine at 116 °C. Removing water by sulphur dioxide and bromine can also achieve complete dehydration, and this also brominates the picoline, lutidine and collidine [Eb. p. 24].

Iodine does not normally require any further treatment unless it is moist. It can be dried over phosphorus pentoxide or resublimed in the presence of calcium oxide.

The quality of the sulphur dioxide is relatively difficult to control. For the preparation of small quantities, it is most practical to use gaseous sulphur dioxide. By evaporating the sulphur dioxide in the container it is purified. When using gaseous sulphur dioxide, it must be remembered that the heat of reaction in the KF solution is higher (i.e. the heat of condensation) and this requires a stronger cooling or a longer time for the reaction. The amount of sulphur dioxide is simple to regulate as the reaction vessel can be placed on a balance.

A freshly prepared KF solution does not have a stable titre and therefore should not be used immediately. It is advisable to wait 1 or 2 days until the initial, i.e. the most severe titre loss has occurred. Then the titre should be determined and the solution can be used for titrations.

The composition of the KF solution is relatively easy to analyse. The active iodine content is determined most practically by a Karl Fischer titration. For example, 10 ml of standard methanol containing 5 mg H_2O/ml (= 50 mg H_2O = 2.778 mmol) are titrated with the KF solution using a standard procedure. Thus the content of iodine can be calculated. This method is applicable to both one-component reagents and the titrant components of two-component reagents. The sulphur dioxide and pyridine contents can be determined by a combined iodometric-alkalimetric titration method. 1 ml of KF solution is pipetted into approximately 20 ml water shaking vigourously. The solution decolourizes. The excess of sulphite is oxidized by addition of a 0.1 mol/

l iodine solution (until a light yellow colouration occurs). The oxidation follows the Bunsen reaction as we are working in aqueous solution.

$$2H_2O + SO_2 + I_2 \rightarrow H_2SO_4 + 2HI, \qquad (4.1)$$

This means that 1 mol SO_2 liberates 4 equivalents of acid which are partially bound by the pyridine present. When this solution is titrated potentiographically with a 0.1 mol/l sodium hydroxide solution, a titration curve with two inflection points is obtained. The first EP is reached when the free acid has been titrated, the second EP represents the reaction:

$$C_5H_5N \cdot HX + NaOH \rightarrow C_5H_5N + NaX + H_2O. \qquad (4.2)$$

The reagent consumption between EP 1 and EP 2 corresponds to the pyridine content and the total consumption (devided by 4) is the original amount of sulphur dioxide. This method of determination is also suitable for two-component reagents. The sulphur dioxide content of the latter can also be determined by a direct iodometric titration in aqueous solution.

4.5 Standardization of Titre

Determining the titre is one of the routine tasks of every laboratory. It is necessary because the titre of an analytical reagent changes or can change. The frequency of titre control can depend on many factors, i.e. the type of reagent used, the equipment used and last but not least the accuracy desired.

Methanolic one-component reagents require a daily titre determination as the change due to self-decomposition can be as high as 1% per day. If a similar reagent, containing 2-methoxyethanol as the solvent is used, a weekly titre determination will usually suffice. Titre determination of two-component reagents is not normally necessary provided moisture from outside does not infiltrate into the titrant component. As this can never be ruled out with absolute certainty, an occasional determination is to be recommended. Determining the titre also serves another pupose – it is a check of the working conditions of titration and a control of the tightness of the equipment being used. A confirmation of previously obtained titre values will give confidence to the analyst in the interpretation of his experimental data.

The degree of accuracy desired is also a significant factor of titre determination and its frequency. If moisture is only to be determined superficially and it is immaterial whether the water content is found to be 0.05% or 0.06%, then a variance of 20% can be accepted. Such a case warrants only a periodic titre check. If, however, water in milk is being determined for example, i.e. water is the main constituent, then a precise titre determination remains unavoidable. In practice, the titre determination is carried out immediately prior to the analysis and serves as a functional control at the same time.

In order to standardize the titre, a substance of known water content, like a hydrate or a known amount of water, is titrated with the KF solution to be used. Unlike in oth-

4.5 Standardization of Titre

er analytical procedures, the titre of a KF solution is not expressed in mol/l but rather in mg H$_2$O per ml of titrant. It is designated Water Equivalent (WE) and is derived from the following formula:

$$WE = \frac{\text{weight of water in mg}}{\text{amount of reagent used in ml}}$$

The determination of the water equivalent should always be carried out under same conditions as the intended water determination. When standard procedures (see Section 3) are normally employed, the same method should also be used for the standardization titration. The type of solvent and the volumes involved should be approximately the same, and the amount of water used for the standardization should be of the same order of magnitude as the amount of water anticipated to be determined. An exception to this rule is the determination of very small amounts of water, as the amount of water for a standardization titration would be so small that the slightest dosage error could influence the results significantly.

4.5.1 Standardization Using Solid Hydrates

Sodium tartrate-2-hydrate has evolved as the primary standard for titre determination. This hydrate has a water content of 15.66%, is very easy to prepare and is available in sufficient purity. It is stable under normal laboratory conditions, does not effloresce and is not hygroscopic even at high humidities. Apart from that, its water content can be determined by an independant method – by drying at 150 °C to constant weight. Thus, nearly all the requirements of a primary standard are fulfilled. However, the solubility is unsatisfactory. Sodium tartrate-2-hydrate dissolves only slowly in methanol and can therefore only be used in a finely powdered form.

Titre standardization is carried out by filling a titration cell with a known amount of methanol or solvent component of a two-component reagent, and dehydrating by pre-titration as usual. 100–1000 mg sodium tartrate-2-hydrate are accurately weighed out and dissolved with stirring. This can take 3–5 min even if it is finely powdered. The solution is then titrated using the KF reagent. A stable end point is normally achieved if the sodium tartrate has completely dissolved.

Other hydrates can also be used for titre standardization. Oxalic acid-2-hydrate is equally suitable. Sodium acetate-3-hydrate is less recommendable as it effloresces slightly when in contact with air. Citric acid monohydrate is somewhat hygroscopic and absorbs moisture at higher humidities. Other hydrates, which can also be mentioned, are ammonium oxalate, ammonium ferrous sulphate, potassium aluminium sulphate, potassium citrate, potassium sodium tartrate, potassium tartrate, sodium citrate, sodium hydrogen tartrate and sulphosalicylic acid [Aqu. p. 116].

4.5.2 Standardization Using Water

Determining the titre using pure water would appear to be the simplest way [7002]. However, the problems of measuring 20–200 µl amounts are often underestimated. This requires an extremely accurate dosing technique. Most commonly used are micro-litre syringes, which achieve a reproducibility of about 1 µl working carefully.

Weighing by difference is also sometimes used to determine the amount of water administered. The tip of the filled syringe is sealed by sticking in a piece of rubber. The syringe is weighed, the water is then dosed and the syringe is sealed and weighed again. Sometimes the titration vessel is weighed as well, though this is only possible for manual titrations. The titration vessel (an Erlenmeyer flask, for example) is filled with solvent, dehydrated by pre-titration, sealed with a ground glass stopper and weighed. 2–3 drops of water are then added and the flask is weighed again.

Bonauguri and Seniga [5502] found the titre determination using pure water to depend on the amount of methanol present. In small amounts of methanol, the water concentration is relatively high and this leads to a higher titre of the KF solution. Based on our present understanding, this would mean that the Bunsen reaction is partially in effect. Eberius and Kowalski [5603] found the dependancies to be complex and could not enlighten the causes.

Summing up, it can be said that the direct administration of water is a good and above all quick method of titre standardization. Its accuracy depends mainly on the dexterity and agility of the technician.

4.5.3 Standardization Using Alcohol-Water-Mixtures

Measuring problems can be lessened if the water is diluted with a suitable solvent. Methanol, the standard solvent normally used for KF titrations, is also used for this purpose. Karl Fischer used a water-in-methanol mixture for standardization of the KF reagents and it is still used today. A titre of 5 mg H_2O per ml methanol is the preferred concentration.

To prepare such a standard, a known amount of water is added to (nearly) water-free methanol. The small initial water content of the methanol should be taken into account. It is most practical to fill a 1 litre measuring flask with methanol out of a freshly opened bottle, thus ensuring a low water content (approximately 0.1 mg per ml = 0.01 %). The measuring flask should have been rinsed previously with methanol as no equipment is really "water-free". After filling the flask, 20.00 ml of this methanol are used for the determination of its water content. This determination does not require high accuracy and is only carried out to evaluate a correction factor, which is normally about 2% of the total water amount. Then exactly 5.00 ml of water are added to the measuring flask and the flask is filled up to the 1 l mark with the same methanol. The methanolic solution thus prepared contains 5.00 + x mg of water per ml, where x represents the initial water content of the methanol. One can also reduce the amount of water added to the methanol so that a solution containing exactly 5.00 mg water per ml is obtained.

The measurement of small amounts of water can be solved this way. The desired amount of water can be dosed using 5, 10 or 20 ml pipettes, i.e. 25, 50 or 100 mg of water. These amounts of water are normally applied to the standardization of Karl Fischer reagents.

But the methanol standards have also disadvantages. Special care is called for by the hygroscopic nature and the volatility of the methanol. The water content increases rapidly in an open bottle [8104], which means that bottles have to be closed immediately after use. Handling with pipettes calls for particular attention. Usually a clean pipette must first be rinsed with the standard-methanol, as there is no guarentee that

4.5 Standardization of Titre

the pipette is absolutely water-free. Then the pipette can be used immediately to measure the desired amount of standard-methanol. Using the pipette a second time after a few minutes also can cause errors, as the methanol residues absorb moisture from the air. Naturally, pipettes should not be filled by suction with the mouth but always using a suction ball. Dispensing the standard methanol by a plunger burette is very practical, and it is the best way to exclude atmospheric moisture. This way ensures a reliable dosage procedure and offers the possibilty to choose each volume.

Recently, solvent mixtures like xylene/butanol have been recommended as water-standards [8104]. Such mixtures are less hygroscopic, less volatile (and not hydrophobic). Thus, the interferences attributed to methanolic solutions are reduced to about one tenth. Riedel-de Haën applied for patents and produces such a solution, named Hydranal Standard.

5 Equipment

The Karl Fischer titration is widely used all over the world, and it is carried out in many different ways and with very differing apparatus and equipment. Automatic titrators are preferred, capable of performing a titration to completion without manual intervention. These titrators have been especially developed to meet the requirements of the KF reaction – a delayed endpoint adjustment and the exlusion of atmospheric moisture. Multiple-purpose instruments like potentiographs are also suitable for KF titrations in certain circumstances. For manual titrations, combinations of more simple equipment are still widely used. Many standards like ISO 760 [7806] and ASTM E 203 [8112] describe the application of proven titration equipment. This section includes a few examples of titration apparatus, from simple glass burettes to microprocessor-controlled automatic equipment. All types of equipment can not be covered.

5.1 Equipment for Volumetric Analysis

5.1.1 Simple Titration Set-Ups

Manually performed titrations are still justified today, above all when a water determination is only occasionally carried out or if the applicability of the Karl Fischer titration is being investigated. For initial trials, the Karl Fischer titration can be carried out using a normal laboratory burette and a 50 ml or 100 ml Erlenmeyer flask. It goes without saying that all equipment used must be dry. Using this equipment, a rapid titration is to be recommended, and slightly over-titrating to ensure a rapid end-point adjustment. The amount of water to be determined should not be too low. An experienced technician can achieve acceptably reliable results this way. Table 5–1 shows the results of a series of titrations carried out in our laboratory. The water content was calculated from the actual titre of the KF titrant, determined in an automated KF titrator. If the titre had been determined manually under the same conditions, the deviation of the average value would probably be less.

If the accuracy and reproducibility are to be improved, the influence of the atmospheric humidity has to be reduced. The Karl Fischer solution should be protected from moisture by using a sealed burette. Such an example is described in ASTM E 203, a titration apparatus according to Pellet. This type of burette can be equipped with an additional dryer, which is necessary for when purging with air. A drying tower or a gas washing bottle, filled with dessicants, are useful.

Table 5-1. Karl Fischer titration using an Erlenmeyer flask and a burette. (Each value averaged from 10 titrations.) Titrant: Hydranal-Composite 5

Water added mg	Water found mg	Standard deviation mg	%
100	100.61	1.53	1.52
50	50.42	1.21	2.22
25	25.30	0.65	2.56
10	10.44	0.32	3.08
5	5.25	0.27	5.30

The volume of air is considerably reduced when this burette is filled by evacuating the upper part of the burette, using a suction ball, for example. Drying tubes also have to be used in this case, to exclude moisture. Plunger burettes, as described in DIN 51 777 [8303] for example, are highly recommended. They can be connected with the suction tube directly to the bottle containing the KF solution, and with the pressure tube to the titration vessel. A few types of plunger burettes are inexpensive and enable a reliable exclusion of atmospheric moisture.

Improvements to the titration vessel are also possible. The neck of the Erlenmeyer flask can be narrowed by inserting a stopper which has only a small opening for the tip of the burette. Also, it is practical to stir the solution by a magnetic stirrer than it is to shake manually. The rate of stirring should be regulated so that no air is mixed into the solution. If a plunger burette is used, flexible tubing can be connected to the Erlenmeyer flask by means of ground glass joints.

The Normag Karl Fischer burette (Fig. 5–1) combines both burette and titration vessel and ensures complete exclusion of moisture, even during titration.

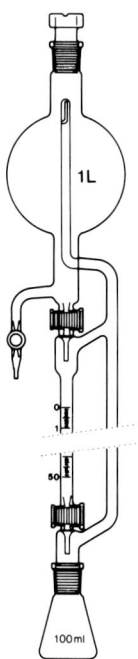

Fig. 5-1 Normag special burette for Karl Fischer titration. (With permission of Otto Fritz GmbH, Hofheim/Ts.)

5.1 Equipment for Volumetric Analysis

A three-neck flask is also suitable as a titration vessel. The burette is connected to one of the three ground glass joints. The second joint is used for addition of solvent and samples. The third connection is fitted to a dry air supply (see Section 6.4) to exclude atmospheric moisture. Many other forms of titration cells are common although they are more specific to particular applications. When constructing or choosing a titration cell, several aspects should be considered. The cell should not be larger than necessary as a larger dead volume contains more moisture. The inside surfaces should be as smooth as possible and preferably be made of glass or PTFE. The inside surface will become wetted with methanol which is hygroscopic and this can cause vanishing end points. Pockets of methanol can be formed in corners and at seals. They absorb and release water thus causing fading end points. Also, the cell should be airtight so it can be vented with dry air if necessary. In many cases, commercially available titration cells are to be preferred as they have been specifically designed to meet the requirements of the KF titration. A few cells have a delivery port for solid substances.

If visual end-point indication is to be excluded as a possible source of error, a double platinum electrode (two platinium wires, each 1–2 mm in diameter and 5–12 mm long), equipped with an electric circuit, can be used. Figure 5–2 shows a simple electric circuit with a galvanometer. A laboratory pH meter can be used for this as it is basically a voltmeter with a high impedance. The double platinium electrode is connected to the pH-meter and then it is polarized. Several pH meters are already fitted with

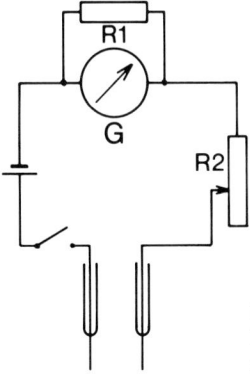

Fig. 5-2 Indication circuit using polarized electrodes according ISO 760 [7806]. $R1 = 10\,k\Omega$, $R2 = 5\,k\Omega$, $R_G = 350\,\Omega$. (With permission of Deutsches Institut für Normung, Berlin)

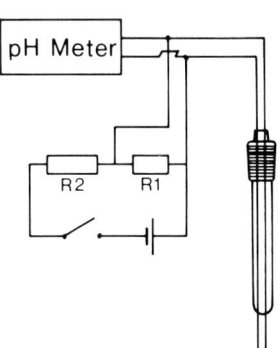

Fig. 5-3 Indication circuit to be used in connection with a pH meter. $R1 = 100\,k\Omega$, $R2 = 200\,k\Omega$

polarization circuits. For other instruments the polarization unit is available as an accessory. A do-it-yourself polarizer, as shown in Fig. 5-3, is also easy to construct. It can supply a polarization current of approximately 5 µA and gives a detection voltage of 20–500 mV. These values can be altered by changing the resistors. Lower resistors increase the polarization current and reduce the responsivity.

5.1.2 Multipurpose Titration Equipment

There are many instruments commercially available which are intended for potentiometric or bipotentiometric titrations or which are suitable to titrate to preset potentials. They are usually referred to as potentiographs or as end-point titrators. It seems to be possible to use these instruments for KF titrations, too. Doing this, the characteristics of the KF titrations have to be taken in consideration, i.e. the sensitivity to atmospheric moisture and the delayed end-point adjustment. This is sometimes not easy to achieve.

The problem of exterior moisture can be eliminated by using a tightly-sealed titration cell. A few titration cells are suitable and can be used. In other cases, cells especially designed for KF titrations are offered by the manufacturers as accessories. In any case a separate cell for KF titration is necessary as switching from titrations in aqueous solutions to KF titrations is out of the question.

Greater problems are caused by the delayed KF reaction. When titrating with a potentiograph at normal speed for example, the dosage can "overtake" the chemical reaction and a noticeable excess of iodine becomes apparent in the solution before the equivalent amount of titrant has been added. Figure 5-4 depicts this behaviour. A very slow attainment of equilibrium is shown by curve A, which represents a one-component reagent containing pyridine as the base. The inflection point, which normally signifies the end point in potentiometric determinations, is far too early and the erratic potential caused by the erratic reagent supply does not permit reliable interpretation of the graph. If a non-pyridinic reagent is used (curves B, C) the attainment of equilibrium is accelerated. A KF titration is most feasible using a two-component reagent (curve C). These graphs cannot be generalized as the deviations – those of reagent A in particular – are caused by many factors. The deviations depend on the composition of the reagent, on the delivery rate of the titrant and on the water content of the sample. The greater the amount of water being titrated, the more accurate is the end-point indication, because the reaction is accelerated by the accumulation of both sulphur dioxide and pyridine.

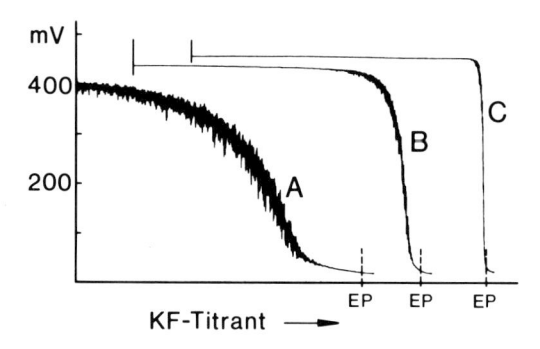

Fig. 5-4 KF titration of 25 mg of water using a potentiograph. Indication circuit according Fig. 5-3, resistors 25 and 50 kΩ. Titration time: 5 min. Reagents: A = pyridine based KF reagent, Riedel-de Haën No. 36115; B = Hydranal-Composite 5; C = Hydranal-Solvent and Hydranal-Titrant. (EP = calculated end points)

Attaining an end point can be improved by titrating to a predetermined potential, for example to 30–50 mV as depicted in the graphs in Fig. 5–4. Here, one titrates to a defined excess of iodine, accelerating thus the end-point attainment and improving the reproducibilty of the end point. End-point titrators are particularly suited to this type of titration technique. If it is also possible to delay the end-point cut-off by 10–30 s for example, the resulting titration equipment then resembles automated KF titration equipment.

Under these conditions, several commercially available potentiographs or end-point titrators can be used for KF titration. An existing instrument should be tried out or its suitability for a particular application can be discussed with the manufacturer. Nevertheless, a polarizer, either built-in as a part of the equipment or added as an accessory, will be necessary. Multiple-purpose tritrators, which are also intended for KF titrations, are also available. The Memo-Titrator DL 40 RC (Mettler) is a microprocessor-controlled instrument which is adapted to the KF titration.

5.1.3 Karl Fischer Titrators

The specific character of the KF titration requires an instrument specific to this application. Many instruments have been developed by a few manufacturers all over the world and they are preferably used in the laboratories to determine water. The models and performances of instruments used today vary greatly. Several types of equipment will be covered in this section.

Simple and inexpensive KF instruments are only little more than an indicating system and a titration vessel with a stirrer or just a connection to an external vessel. These instruments require manual administration of the titrant using a burette. Automated titration equipment merely extends this basic principle. The reagent is added automatically until the end point is reached, and the delivery rate is adjusted according to the water content in the sample. The end point can be preset, and the stability can be checked by setting a persisting time, thus delaying the cut-off of the titration. Newly developed instruments change the working medium automatically. Burettes are refilled automatically, and the titre of the reagent can be standardized by a calibration titration. The results are displayed digitally, printed out and calculated in different ways.

The 633 Karl Fischer Titrator made by Metrohm Ltd. is universally applicable for water determinations in liquids, solids and gases. Its control program is adapted to the characteristics of the KF reaction with special regard to the kinetic behaviour of the new KF reagents. The end point is detected by means of controlled-current potentiometry applying a double platinum electrode. Routine titrations with the 633 KF Titrator require but one keystroke. Delay time can either be selected between 2 and 900 s or set to "∞" for conditioning (drying) of the solvent. The extraction time for back-titrations is adjustable between 20 s and 90 min.

The reagent-dosing equipment available for the 633 KF Titrator ranges from low-cost burettes to microprocessor-controlled Dosimats. The instrument can thus be adapted to a wide variety of uses and budgets. The titrant volume is displayed digitally. The variant employing a microprocessor-controlled Dosimat can be linked to a desktop calculator for automatic result evaluation.

A number of practical accessories are available: Septum stopper for injecting liquid samples; weighing spoon with protective tube and titration vessel with lateral

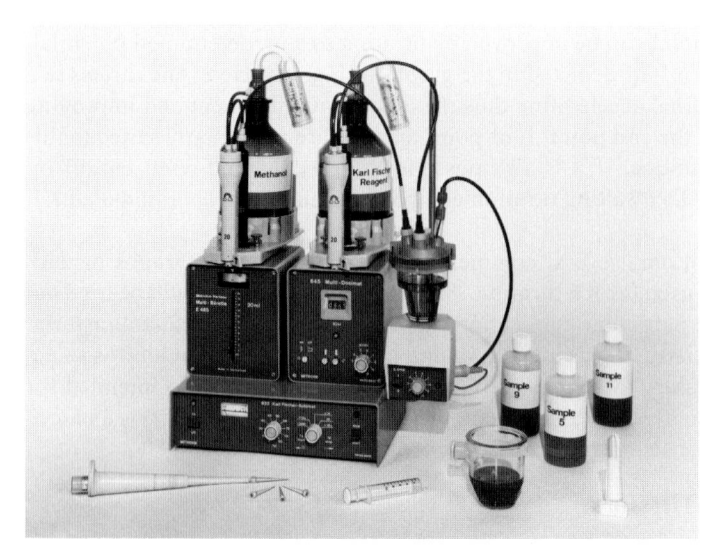

Fig. 5-5 The 633 KF Titrator from Metrohm equipped with 485 Manual Burette and 645 Motor Burette

ground-joint aperture for introducing solids; water determination in gases by means of solenoid valve and gas meter. Figure 5–5 shows a low-cost variant of the 633 KF Titrator with manual solvent addition.

The newly developed 658 KF Processor combines the excellent features of the 633 KF Titrator with microprocessor intelligence. The steps involved in the various titration procedures are carried out automatically in the correct sequence. The program controls and checks all the instrument's functions: Direct titration, back-titration, conditioning of reagents, titre determination, evaluation of result, calculation of mean value and standard deviation (elimination of outliers is also possible).

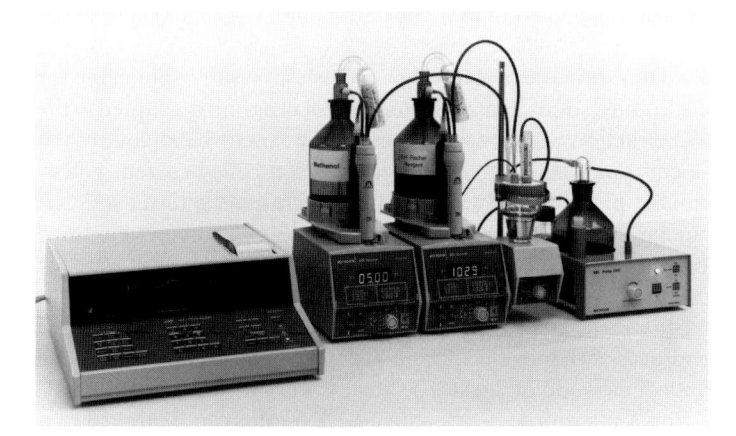

Fig. 5-6 The 658 KF Processor from Metrohm equipped with two 655 Dosimats for direct and back-titrations and 661 Pump Unit (right). The basic unit consists of 658 KF Processor, titration cell and one 655 Dosimat

5.1 Equipment for Volumetric Analysis

Equipped with optional interfaces for balance and external data system the 658 KF Processor represents an instrument well suited for integration into the modern analytical laboratory. The newly developed titration cell eliminates the detrimental effect of ambient humidity. The optional new 661 Pump Unit allows easy renewal of reagents. If the water content of solid samples is determined by means of the 613 Drying Oven, the Pump Unit is used to generate the dry air stream which transfers the water given off by the heated sample to the titration cell.

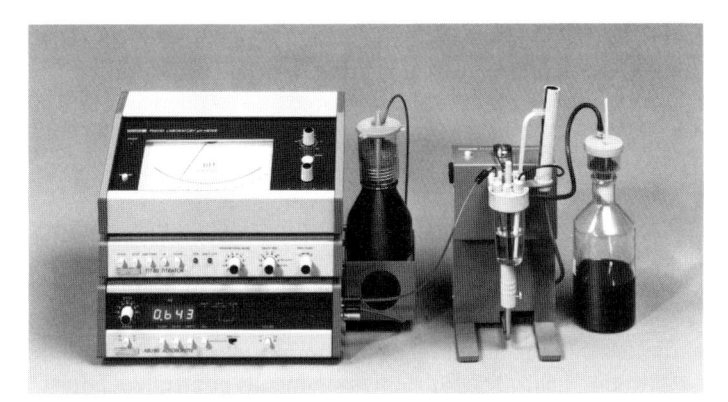

Fig. 5-7 The Karl Fischer Titration System ETS 850 from Radiometer, Copenhagen

Radiometer Copenhagen manufactures the Karl Fischer Titration System ETS850 for the determination of water in solids and liquids (Fig. 5-7). It is easy to operate because it titrates automatically, and the burette is refilled automatically. The reagent consumption is indicated directly by a four-digit display. The automatic shut-off delay can be set stepwise to delay-times between 2 and 50 seconds. Thus, titrations with high reaction rates can be carried out quickly, and slowly reacting samples can be titrated to completion. The polarization voltage is adjustable, and standard resistors can be changed, thus ensuring optimum titration conditions for every application, even when the solvent has a low conductivity. Plunger burettes with volumes between 0.25 and 25 ml permit accurate determination of both large and small amounts of wa-

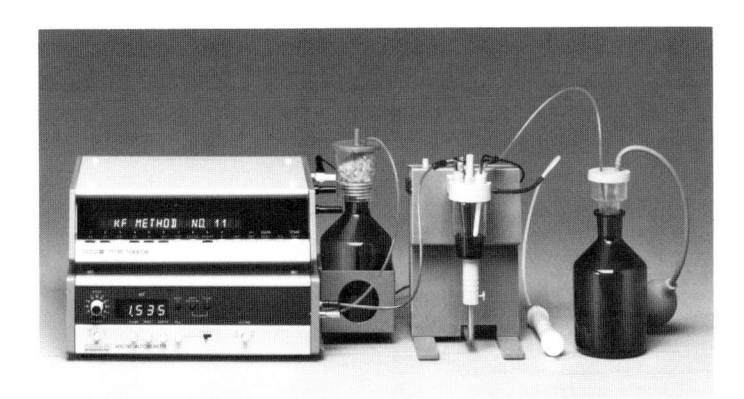

Fig. 5-8 The Multi-Titration System MTS 855 from Radiometer, Copenhagen

ter. The titration vessel and the burette are sealed to prevent errors due to atmospheric humidity. The titrant in the stock bottle and in the burette is also protected against light.

The MTS855 Multi-Titration System from Radiometer is a microprocessor-controlled system for Karl Fischer titrations (Fig. 5-8). It offers perfect titration control thanks to the automatic calculation of the size of each increment of KF reagent. Dynamic monitoring of the electrode signal fluctuations ensures that the doses are appropriately decreased as the end point is approached. The delay time is adjustable between 1 and 999 s. The water content is automatically calculated in user-selected units, such as ppm, %, mg/l, etc. The settings, calculation data and unit for titration of up to 31 different samples types can be individually stored and recalled at the push of a button.

The new Karl Fischer Titrator DL 18, offered by Mettler Instrumente AG, Greifensee, Switzerland, allows for moisture determination in the range from a few ppm to 100% water content. The microprocessor-operated instrument combines burette, sealed titration vessel, magnetic stirrer and a pump system for changing solvent in one compact unit (Fig. 5-9). An appropriate design prevents leaking of outside moisture into the titration vessel and all necessary air inlets are sealed with drying tubes. Liquid samples can be injected through a septum, solid samples are inserted through a delivery opening in the titration head. The titration is kinetically controlled, adjusting speed of titrant delivery to the expected content of water. Thus, a large determination range is achieved without a loss of titration speed or precision. Typically 50 mg of

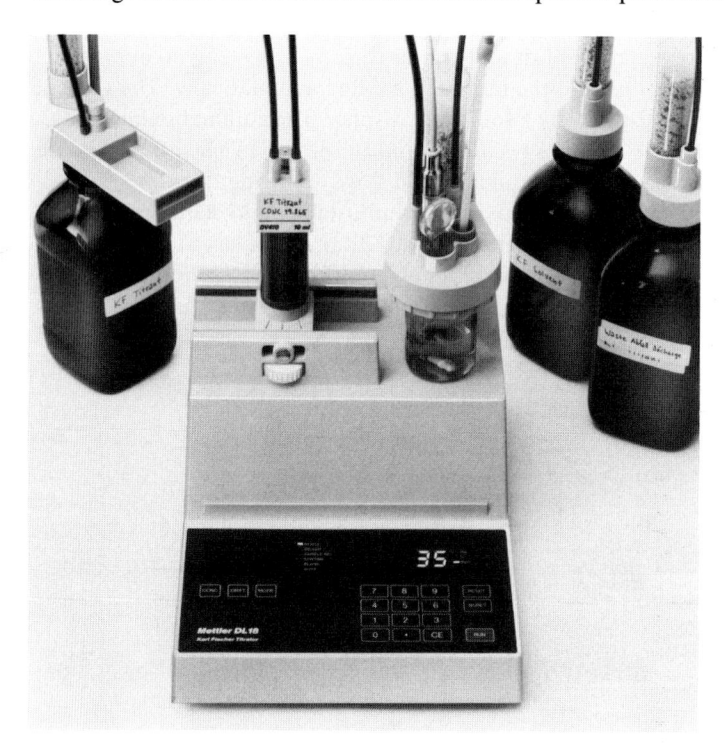

Fig. 5-9 The DL 18 Karl Fischer Titrator from Mettler, Switzerland

5.1 Equipment for Volumetric Analysis

water are titrated in one minute with high precision. An automatic correction for background water consumption further enhances the results.

The user communicates with the instrument via a keyboard and display. Results are calculated in mg, ppm or % water and may be reported on a printer. A balance can be connected directly, and a computer by an optional interface.

The titrator can be used with any of the commercially available Karl Fischer reagents, as the instrument takes care of the different kinetic behaviour of the various reagents itself.

Schott-Geräte supplies the Dead-Stop Titrators TR 151 and TR 152 for Karl Fischer determination. Both instruments can be set in different ways to optimize the working conditions: the delivery rate of the burette, the cut-off current and the delay time are adjustable. They also can be used for back-titration. The end point is measured biamperometrically (polarization voltage 100 mV). The cut-off current can be preset in six steps between 1 and 50 µA. The indication current is indicated on a large analogue display with high resolution and the course of titration can be followed very closely. The titration speed can be varied from 1–20 min for one burette volume. For KF titration the delivery rate reduces automatically as the equivalence point is neared. The delay time can be set to one of seven ranges between 2 and 100 s, and the position "∞" is for continuous dehydration of the titration vessel. Both instruments are equipped with a 37 pin plug for data processing and data systems. The titrator TR 151 is intended for a single burette. The titrator TR 152 can be equipped with a second burette for dispensing of solvents. Several titration vessels are available. For the titration of liquid and gaseous samples, a cell made completely out of glass and fitted with a septum can be used. This reliably excludes atmospheric moisture. For solid samples a solid lock made of PVDF can be introduced. The cell can be emptied by means of a valve at the bottom. Further accessories include burette tips which inhibits back-diffusion and vessels with water jacket and removable hoods.

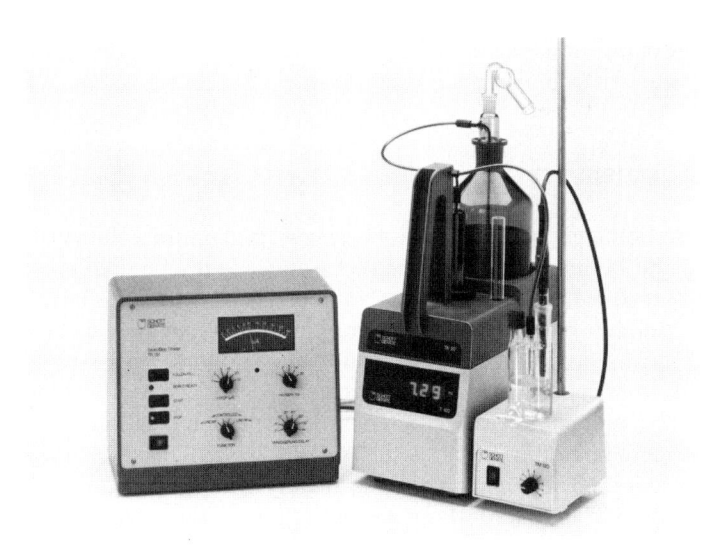

Fig. 5-10 The Dead-Stop Titrator TR 151 from Schott-Geräte, Germany

Fisher Scientific manufactures the "Automatic K-F Titrimeter System". It titrates automatically, but the important working parameters can be preset in order to ensure optimum titrations. The end point is indicated biamperometrically and the indication current is displayed on a large-scaled microammeter on the front panel. The cut-off current can be pre-selected between 0 and 50 µA. During the titration the addition of the reagent is regulated dynamically. First the titrant is measured rapidly. At selected proximity to the end point, the titrant delivery rate tapers off continuously to a preset minimum. Both the proportioning band and the minimum delivery rate can be adjusted to adapt the working conditions to the requirements of the sample, the working medium or the reagent. An adjustable delay timer (0–100 s) confirms end-point persistance. The volume of reagent consumed is displayed on a LED readout. The titration vessel has a large capacity of about 400 ml. Thus several samples can be analysed successively in the same solvent. Solvent delivery and draining of cumulative waste from the titration vessel are accomplished by pushbuttons for actuating a built-in air pump and solenoid valves. A waste button also allows purging with dry air to minimize moisture initiation. Vents are fitted with drying tubes to limit moisture entry during operation. Liquid samples are injected through a septum, solid samples are introduced by means of a screw cap. Gases can be analysed by introducing a gas inlet tube.

The AQV-5 Aquacounter from Hiranuma (Japan) is an instrument for the automatic determination of water. It is designed to measure water contents of low concentrations with high accuracy in a short time. The instrument contains a microcomputer thus enabling to automate the titre determination, back-titration, calculation of the results and statistics. The result is displayed on a 5-digit display in ml titrant, mg of water, percent or ppm. It can be printed out on the standard printer along with the information entered before, like: sample number, sample volume, moisture concentration in percent or ppm, average value, standard deviation. With an automatic electronic balance used in conjunction with the titrator, sample weighing can be automated, making it possible to reduce work and enhance accuracy in analysis. The instrument is equipped with two burettes for direct titration and back-titration. The working parameters, like delay time or titration time, are preset using the keyboard. It is designed like the coulometric instrument from Hiranuma depicted in Fig. 5–12.

5.2 Equipment for Coulometry

For coulometric KF titration a greater technicality is necessary than for volumetric determination. It seems to be impossible to construct suitable instrumentation by oneself. Commercially produced instruments first had to become available to introduce this analytical method into the chemical laboratory. It is therefore understandable that this technique has only acquired practical significance in the last few years.
The development of KF coulometry began about 25 years ago. The first results were published by Oehme [5801] and by Meyer and Boyd [5901]. Other publications deal with the construction of suitable coulometric cells [6301, 6702, 7401, 7501] which have to meet particular requirements. The cells have to be absolutely tight as micrograms of water are to be determined in a laboratory whose air space contains kilograms of water. The cathode reaction must be separated from the anode reaction in order the prevent the formation of reducing substances in the anode compartment, which would

5.2 Equipment for Coulometry

Fig. 5-11 The 652 KF Coulometer from Metrohm, a microprocessor-controlled instrument for determining very low water contents

behave like water as they consume iodine. As the electric resistance has to be kept within limits, the choice of a suitable diphragm was a decisive factor. Along with these developments, appropriate circuitry was also developed [7406] and instruments available today are "push-button" operated.

The first commercial instrument was, to the best of the author's knowledge, the "Aquatest" of the Photovolt Corporation in New York. This instrument has an automatic drift correction which eliminates errors due to atmospheric moisture by calculation.

Mitsubishi (Japan) manufactures the Moisture Meter CA 02 which features a sealed cell made completely of glass, thereby ensuring a low drift. The results are displayed digitally in μg of water.

The 652 KF Coulometer manufactured by Metrohm Ltd. is shown in Fig. 5-11. This instrument applies pulsed current coulometry and is especially well suited for determining water amounts ranging from 10 μg–10 mg. The air-tight cell combined with drift compensation allows to take full advantage of the high sensitivity of the coulometric method. The cell is conditioned automatically as soon as the instrument is switched on. Operating is extremely simple. To start a water determination, only the "GO" key has to be pressed, and the sample has to be added. The instrument's microprocessor controls the analytical procedure and displays the result in μg of water. If the mass of the sample is entered, the result is displayed as a mass fraction (parts per thousand or ppm). When working with dissolved solid samples the blank value of the solvent can be entered and is taken into account for result evaluation. Extraction times can be set and a drying oven for determining the water content of solid samples is available.

The Aquacounter AQ-5 from Hiranuma (Japan) is designed for automatic coulometric water determination. It is depicted in Fig. 5-12. All operations, such as titration control, end-point detection, calculation of the results are carried out by a microprocessor. The results are indicated on a five-digit digital display. If data are entered by the function key, the input data are displayed, too. All data can be printed out on the

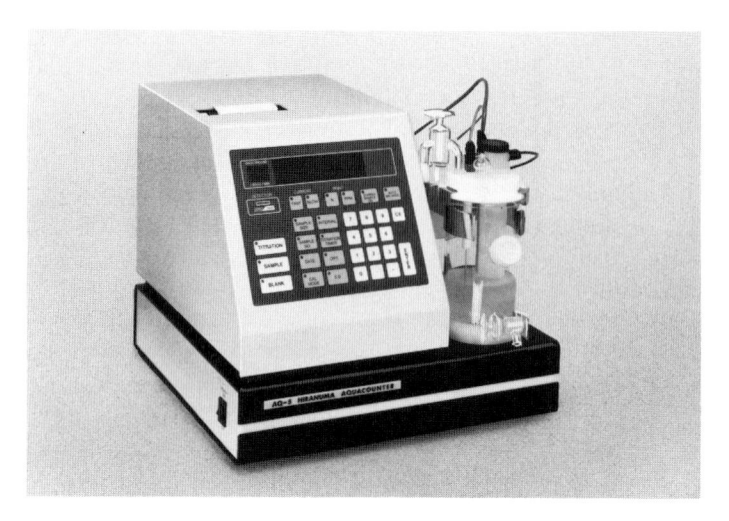

Fig. 5-12 The AQ-5 Aquacounter for coulometric determination of water from Hiranuma, Japan

printer. The instrument is operated by the panel keyboard, which is of the sealed structure resistant to chemicals. The coulometric cell of the Aquacounter AQ-5 features an ion exchange membrane instead of a diaphragm to separate anodic and cathodic compartments.

Fisher Scientific's Coulomatic Karl Fischer Titrimeter (Fig. 5-13) also utilizes microprocessor control to automatically determine sample moisture with an accuracy of 5 µg in the range 10–1000 µg water; 0.5% in the range 1–10 mg water. A six-digit display shows results directly in micrograms, milligrams, percent, or parts per million, automatically corrected for drift or blank. A maximum titration rate of 2500 µg water per minute, coupled with automatic tapering of the rate by proportioning in the vicinity of the end point, enhances analytical speed with no sacrifice of accuracy. An alphanumeric display provides clear prompting for data entry and method setup, as well as status and diagnostic messages. The display and a warning beeper signal when reagent depletion threatens performance. Extraction time, persistence interval, and digitally controlled stir rates are operator programmed. The system interfaces directly with several digital balances and printers.

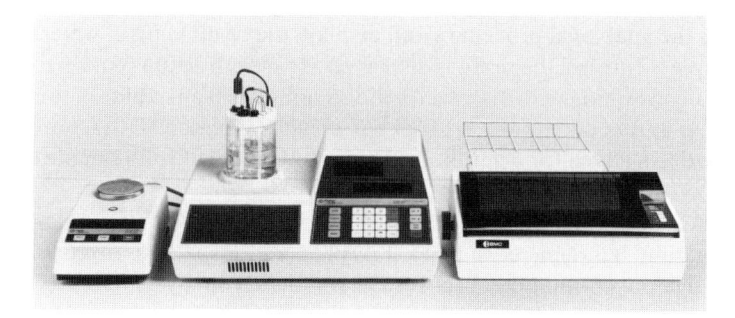

Fig. 5-13 The Coulomatic Karl Fischer Titrimeter from Fisher Scientific

6 Application and Sources of Error

In practice, the water to be determined is not present as free moisture but rather combined as a part of another substance. Thus the water is mostly determined in the presence of an excess of another substance. The determination is only possible when this matrix substance does not interact or when a forseeable interference can be avoided by modifying the working conditions.

During the last half century, the application of the KF titration to many different substances has been intensively investigated and many publications report the results. As the variety of substances are manyfold indeed – ranging from organic and inorganic chemicals to food, naturally occurring substances and technical products – the publications are widely distributed amongst the faculty-specific literature and are often very difficult to learn of or obtain. We have tried to collect as many individual publications as possible. Those which appeared to be of particular interest we have reported in Sections 7, 8, 9 and 10 and referenced in the literature appendix.

Evaluating a publication it should be taken in consideration that the investigations were often carried out under very specific conditions and with very specific goals. The results sometimes have to be evaluated critically. A few examples should make this clear. Some publications conclude a determination as being "feasible" if a visual end point can be detected. In contrast to this, earlier determinations were designated as "not feasible" when a visual end point could not be found. In the light of electrometric end-point indication, the latter is certainly no longer true, today. On the other hand, a "feasible" determination should also remain questionable as long as the reported end point has not been verified to represent the real water content. Later studies therefore often use comparative methods of determining the water content. Now, the question of their accuracy has to be answered, (most of the comparative determinations are less accurate than the KF titration) and also which tolerances can be accepted. The determination of water in a beverage for example, which mainly consists of water, can surely need a reproducibility of 0.5%, whereas for an organic liquid containing 0.01% of water, a deviation of as much as 10% could be tolerated.

Together with this, it should be mentioned that very differing working techniques are employed. In earlier studies, the water content was directly titrated in different liquids without the now customary pre-titration of an alcoholic solvent as described in Section 3. Further differences arise from the types of reagent used. A methanolic reagent is rather less suitable for ketones for example than is a reagent containing 2-methoxyethanol which is much better. As the equipment used can also be different, the results are really only comparable when they are evaluated in connexion with the method, the reagent and the equipment used.

We have therefore declined to judge the studies of applicability and refer the reader to the original publication. Only in exceptional cases we point out a potentially misleading interpretation.

The literature contains also several general rules which are the basic requirements for achieving reliable results and many informations on different interferences which can have a negative influence on the reliability, thus introducing errors into the analysis.

The sources of error are of very differing type. First the substance under investigation can be insoluble in the KF system, including the water, and inhibiting the determination. This often occurs with salts which do not form hydrates (sodium chloride for example) and which include small amounts of water in their crystal structures (about 0.1%). Water of hydration can be determined more easily as many hydrates are dehydrated in the KF system without dissolving, though not always completely and often slowly. The formation of water is also an effect which is observed. Some acids esterify and aldehydes can form acetals, thus producing water. Other reactive components in the KF system can also cause side reactions. Iodine can be added to double bonds or be reduced by reducing agents. Oxidation can occur by reaction with the sulphite or iodide, which is formed during titration. Substances, either acidic or alkalinic, can affect the buffer system of the KF solution and alter the pH, inhibiting or delaying thus the titration or changing the stoichiometry of the reaction. Even hydrocarbons, which are chemically inert, can affect the KF reaction by reducing the alcohol concentration. In some cases, the influence can be so far-reaching as to alter the stoichiomety of the KF reaction.

If the requirements are not fulfilled or not completely fulfilled, i.e. distributing effects could take place, the working conditions have to be modified. There are a number of ways of doing this. A review of general methods for this is presented in this section. Specific effects peculiar to particular substances or classes of compounds will be described in context with those products covered. Furthermore, disturbances caused by external interference, such as atmospheric moisture, temperature variations or titration equipment are also possible. These, as well as indication problems and sample handling will also be covered in this section.

6.1 Working Medium

When carrying out KF titrations, the most important decision is the choice of a suitable working medium, i.e. a suitable solvent. The working medium determines the course of the reaction and must fulfill three fundamental requirements:

1. The stoichiometry of the KF reaction must be assured. According to Eq. (2.22), this requires the quantitative transfer of sulphur dioxide into the alkyl ester of the sulphurous acid.
2. The solvent must be capable of dissolving the sample and the products of the KF reaction.
3. Indication must be possible.

6.1 Working Medium

Methanol is the preferred working medium for KF titrations as it fulfills virtually all requirements. The KF reaction is stoichiometric and takes place efficiently in this medium. Methanol dissolves most components and reaction products of the KF reaction so that precipitation does not occur which could otherwise coat the electrodes and interfere with the indication system. Also, the solubility of most of the samples investigated is relatively good. Most substances are dissolved rapidly by methanol and hence the titration times are comparatively short. The relatively good conductivity of a methanol electrolytic solution enables both a sensitive as well as reproducible endpoint indication either instrumentally or visually.

It is therefore understandable that methanol has become the standard solvent for KF titrations. The working techniques have been developed primarily using methanol as the solvent, and based on its properties, the instrumentation also has been designed and developed for working in methanolic media.

Methanol is limited though in its applications. Its solubility of long-chain hydrocarbons is as equally unsatisfactory as are some side reactions which take place with certain amines, mineral acids and ketones. Other solvents are therefore used in certain circumstances instead of methanol or preferably as mixtures with methanol. When two-component reagents are used, the other solvents can be mixed with the solvent component of the reagent.

Long-chain hydrocarbons are better soluble in 1-propanol and this is the preferred solvent used for their investigation, either pure or mixed with methanol. 2-Propanol has certain drawbacks as the stoichiometry can be shifted slightly. We found deviations of up to 3%. We also know from our investigations reported in Section 2.3 that the formation of an alkyl sulphite is incomplete in 2-propanol.

1-Butanol can equally be used although its high vicosity can pose problems for its application as a solvent. 2-Butanol is less recommendable as the titre deviations are approximately double of those when using 2-propanol. Tert-butanol is even less suitable because as well as a change in the stoichiometry, indication problems and difficulties of mixing have been observed.

Because of its high boiling point, ethylene glycol is used to absorb the moisture in gases and it was thus introduced into the field of KF titration. The KF reaction remains stoichiometric in this medium. However, it must be mentioned that the high viscosity requires a long time for mixing the solvent in the titration vessel after addition of the KF reagent. Ethylene glycol is therefore often mixed with methanol, pyridine or with other solvents.

2-Methoxyethanol (methyl glycol) is recommended as the working medium for the investigation of ketones, as it inhibits the formation of ketals sufficiently to enable a water determination. The KF reaction proceeds stoichiometrically so that no titre deviation occurs. When using certain types of titration equipment, the titration can sometimes be delayed, however. 2-Ethoxyethanol acts in a similar manner. Both solvents require particular care when handling as they can be injurious to health (Section 6.7).

The evaluation of non-alcoholic solvents appears to be somewhat more indefinite. As they can not form esters with the sulphur dioxide according to Eq. (2.22), they should not be considered suitable as working media. This statement is true for many solvents. In benzene, xylene, acetic acid, ethyl acetate, chloroform, dioxane and many other solvents, the stoichiometry of the KF reaction is actually altered. Apart from this, they prove unsuitable as working media as they do not dissolve the products of the KF reaction satisfactorily. There are a few exceptions which are used in practice but can not be plausibly explained by theory.

Pyridine is a proven solvent for investigating ketones and aldehydes. In this working medium, the side reactions of the carbonyl groups are sufficiently suppressed so that a limited water determination is possible. KF reagents with a low content of methanol are preferred for such titrations (see Table 4–2), and it can not be ruled out that the low alcohol content of the titrant assures the stoichiometry of the KF reaction. Pyridine, being polar, sometimes also serves as the solvent for other substances. Because of its basicity, it is used to neutralize sulphuric acid and other strong acids which would otherwise quickly esterify the methanol.

Chloroform is a proven dissolver of long-chain substances and fats and is used for their investigation. It is preferentially mixed with methanol or with methanolic solvent components. A mixing ratio of 1:1 is to be recommended. In exceptional cases, the chloroform content can be increased to as much as 75% although this can cause slight alterations in the stoichiometry, i.e. the titre of the titrant component appears to increase.

The solubility of polar substances is improved by using formamide and this solvent is added to the working medium when protein products and sugars are being investigated. It can also be used at high temperatures. Formamide accelerates the course of the titration [7802], the end point is attained very rapidly and this reduces titration times. The stoichiometry, however, appears uncertain. We found titre deviations of up to 15% for volumetric titrations using pure formamide, decreasing proportionally by successive additions of methanol. In practice, we consider a methanol content of at least 40% to be essential. Contrary to this, Swenson and Keyworth [6301] and Bizot [6702] use pure formamide instead of methanol to prepare coulometric KF reagents (see Section 4.2), which would presuppose a stoichiometric reaction. Safety measures for handling formamide are described in Section 6.7.

Dimethylformamide is sometimes used as a solvent for the investigation of polymers. Klimova, Sherman and L'vov [6703] recommend DMF as the working medium for the investigation of aldehydes and ketones.

Provided the above requirements are fulfilled, the working medium of a KF titration can be varied to a limited extent eliminating thus a few interferences of the KF titration. Changing from methanol to another solvent should be carefully evaluated, however, as the sample to be investigated also becomes part of the system and the stoichiometry could be altered. When trying out or using a new solvent formulation or a new matrix substance, the recovery rate of water should be tested. The easiest way to do this is to titrate the sample and a known amount of water successively in the same solution. Whether a methanolic water-standard is practical for this must be decided in each case.

6.2 The pH

The Karl Fischer reaction can only take place in a certain pH range. According to Verhoef and Barendrecht [7602, 7703] a constant reaction rate is attained between a pH of 5.5 and 8 (see Section 2). In acid solution the rate constant K is lower and the pK

6.2 The pH

value decreases proportionally with the pH. The end-point attainment becomes sluggish or an end point will not be reached at all. For all practical purposes, a pH of 2 is the limit for KF titrations. Titration is preferentially carried out within a pH range of 4–7; using visual indication, this is restricted to pH 6–7. In alkaline solutions side reactions take place. They consume iodine and KF titrations are not possible.

Pyridine buffered reagents have a pH 4–5 and are thus outside the optimum pH range. This is the reason for the well-known sluggish titrations observed with the "classic" KF reagents. Presumably the "catalytic" action of N-ethylpiperidine, which Archer and Jeater observed [6501] is caused by increasing the pH of the KF system. The new pyridine-free reagents optimize the pH thus assuring higher titration rates.

Acids are formed during the titration of water. They are neutralized by the base present in the reagent (see Eq. 2.22). A good KF reagent is normally adequately buffered by a sufficient excess of base contained in the reagent to stabilize the pH during titration.

This buffer equilibrium will be influenced if strong acids or bases are brought into the system, i.e. when water is to be determined in these products. The degree of disturbance caused is difficult to ascertain. It depends on the amounts of acids and their strengths. The amounts of acids which are formed during the pre-dehydration of the solvent and the water determination must also be taken into consideration. Using two-component reagents the buffer capacity of the amount of solvent component can be calculated, so that the total amount of acid can be kept within acceptable limits. If a one-component reagent is used, the solvent selected normally does not have any buffer capacity. It is then difficult to know whether the reagent contains a sufficiently high amount of base to neutralize the acid and to achieve an end point.

It proves most practical to add a suitable base to the solvent at the pre-titration stage. During the pre-dehydration, this base is also dehydrated and a working medium for the investigation of acids is thus obtained. Only moderately strong bases like pyridine are suitable. Strong bases would turn the KF system alkaline thus making the pre-titration impossible. Pyridine-free buffer solutions based on imidazole are also common and ensure an optimum pH range [Hyd. p. 30–32].

The anologous is true of the investigation of bases. Strong aliphatic amines turn the KF system alkaline preventing thus the water determination. They must be neutralized by adding acid before the KF titration is carried out. This is best done in practice by adding a sufficient amount of a weak acid to the solvent prior to pre-dehydrating. The acid is thus also dehydrated during pre-titration and hence a working medium for the determination of bases is obtained. The literature usually recommends acetic acid for this. We prefer to use either propionic acid as it esterifies more slowly, or better still benzoic acid which does not esterify at all and produces very stable end points.

Using a two-component reagent, the solvent component is treated in a similar fashion with either a weak base or a weak acid and then dehydrated by the pre-titration.

A sluggish or a vanishing end point can be an indication of a wrong pH value for the system. In such a case, the pH should be measured. A normal glass pH electrode calibrated in aqueous solution, can be used for this. The accuracy is sufficient for this purpose. After the pH has been measured, the KF titration can not of course be carried out as the glass electrode will have introduced too much moisture into the system.

6.3 Side Reactions

The Karl Fischer solution contains a few reactive components. Iodine, sulphur dioxide, pyridine and methanol are the initial substances. A pyridinium salt of methyl sulphurous acid is formed from this. During the water titration, methyl sulphuric acid and hydroiodic acid are liberated, also in the form of their pyridinium salts. The KF solution behaves like a weak acid. It would be expected that these reactive components would cause many side reactions with the products under investigation. But fortunately this is not the case. Side reactions remain rather the exception.

Strong reducing agents like stannous salts or ascorbic acid reduce the iodine in the KF solution. Reduction by dithionite, which is a quantitative reaction in aqueous solutions, does not take place in the KF system, which means that its water content can be titrated without any difficulty. Additive reaction of iodine to double bonds take place only in exceptional circumstances. Mercaptans are quantitavely oxidized to disulphides.

Strong oxidizing agents like chlorine, nitrous oxide or dichromate can oxidize the iodide to iodine or the sulphite to sulphate.

Carbonates, hydroxides and certain oxides are neutralized by the pyridinium salts and liberate water. These reactions are stoichiometric in many cases.

Several unwanted side reactions also take place as a result of the alcohol components. Certain carboxylic acids are esterified by the methanol, and the majority of aldehydes and ketones produce acetals and ketals. The formation of the water thus produced can be so rapid with many compounds that an end point during KF titration can not be reached.

A few amines, 1,2-diamine in particular, also give fading end points. The reason for this seems to be unknown. Possibly, these compounds are N-methylated, thus liberating water.

There are a number of possibilities to inhibit these side reactions or at least to eliminate their detrimental effects. Reducing agents like ascorbic acid can be oxidized by iodine prior to the KF titration. Oxidizing agents like chlorine or peroxides can be reduced by sulphur dioxide. Mercaptans can be added to double bonds and reacted to sulphides which won't interfere. The formation of water by carbonates and hydroxides can be eliminated in the calculations. The esterification of mineral acids can be sufficiently delayed if 2-methoxyethanol is used instead of methanol. Interference caused by ketones and amines can similarly be suppressed. The formation of acetals can be prevented by cooling to 0 °C so that the water of the majority of aldehydes can be titrated. In certain cases, the water can be separated from a detrimental matrix by evaporation. Using special equipment, it is carried by a stream of dry gas into the KF cell followed by a KF titration.

These examples highlight the many possibilities of overcoming undesirable side reactions. Further details and other variations specific to particular substances are described in Sections 7–10. There it will be shown that most of the difficulties can be overcome and that only a few "problem cases" remain.

6.4 Atmospheric Moisture

Due to the ubiquitous presence of water, moisture is the largest source of error in KF titrations. Moisture can either penetrate into the titration vessel directly from the air or it can be carried by the apparatus used, by the solvent or by auxiliary reagents and equipment.

Moist apparatus does relatively little harm as the error thereby introduced is restricted to one determination, as it is the case of a moist pipette having been used. The error is recognized immediately and can be eliminated by drying all the pipettes for example or rinsing them with the solution to be pipetted before use. The same applies to measuring flasks and to a certain extent to the titration vessel itself.

Air humidity causes concealed errors which are much more difficult to ascertain. The hygroscopic nature of methanol and of other solvents used for KF titrations is basically responsible for this. They are the cause of a continual absorption of moisture by all equipment and solutions used. A pipette which is still moist with methanol can not be reused after a few minutes without first being rinsed. Methanolic solutions absorb moisture from the atmosphere so that stock bottles and measuring flasks have to be immediately resealed after taking a part of the solution. All KF reagents, solvents etc. stored in stock bottles which are connected to burettes should always be protected from exterior moisture by means of efficient and sufficiently large drying fixtures. Plastic tubing used to transfer liquids to and from the burettes is not always impervious to water vapour. If a burette has not been in use for a long time, it is often advisable to discard the first fill. The loss of the expensive KF solution works out cheaper than a wrong analysis or an incorrect titre determination.

The biggest problems are encountered with atmospheric moisture in the titration vessel, often in commercially available titration vessels as well. These vessels often consist of many parts and have different components like electrodes and introductory tubes. The inner surface area is thereby increased and pockets are created in corners and at stoppers and seals which collect methanol. When the titration cell is opened to introduce the sample, or when the spent solution is emptied or sucked out, air will enter the cell from the outside. As 1 litre of air can contain as much as 20 mg of water, several milligrams of water can be brought into the cell this way. The water is slowly absorbed by all methanol-moist surfaces, the titration solution, the inner walls, and pockets of liquid. Thus a potential stock of moisture is created in the cell. When the working medium has been titrated, the water desorbs from the wall and causes a sluggish or vanishing end point even though the cell is tightly sealed. When a drop of liquid runs unforseen out of a pocket or down from the inner wall into the titration solution, a single analysis can be falsified without being noticed. This is one explanation for the occasion observation in KF laboratories that "results obtained in the mornings are more reliable".

To exclude this source of error, Muroi and Ogawa [6302] put the whole KF apparatus into a nitrogen-filled bag and additionally purged the titration cell with dry nitrogen. Drying cabinets, similar to glove boxes, already had been used. Such measures appear very cumbersome and not really necessary for routine laboratory practices as there are a number of simpler possibilities which can be applied either singularly or together:

1. A titrant with a higher WE (5mg H_2O/ml) and a sample as large as possible should be used in order to reduce the percental error.

2. A quick KF reagent should be used to get a short titration time, as the deviation due to external moisture is proportional to time.
3. The titration cell should be tightly sealed, preferably with ground glass stoppers, and should consist of as few parts as possible and be well dried by pre-titration.
4. Working daily, the titration cell should be dried overnight by adding an excess of KF solution in the evening. The moisture absorbed is thereby eliminated. A newly assembled cell should be conditioned in a similar way for at least one hour.
5. If possible, the cell should not be dismantled or washed out. The spent solution should be pumped off, preferably by using a piston pump or a polythene washing bottle. A water jet pump is less recommendable, as it tends to empty the cell rather too thoroughly whereby too much air is automatically carried through the titration vessel.
6. When emptying the cell, the fresh air should be passed into the cell by way of a suitable drying fixture. This requires that the pump fixture be connected to the cell by means of a tightly sealed stopper.
7. The safest method is to rinse the cell with dried air when it is opened to add solvents or to weigh out solid samples. An arrangement for this is shown in Fig. 6–1. Fresh air is introduced by a flow-meter R at a rate of 300 ml per minute. The air normally flows through the gas washing bottle W 1 away to the outside. As this wash bottle is filled to approximately 12 cm with concentrated sulphuric acid, it maintains a slight pressure in the titration cell Z. Opening the cell to introduce a sample, the air flows through the wash bottle W 2 and the drying tower F into the titration cell Z and from there through the sample delivery port to the outside. The infiltration of non-dried air is prevented this way. When the delivery port is closed again, an internal pressure is restored and this safeguards the cell against the diffusion of moisture. If the cell is leaking somewhere, the wash bottle W 2 will indicate this by forming bubbles. This arrangement is a reliable moisture safeguard. Nitrogen can equally be used instead of compressed air, in which case the drying tower is not absolutely necessary. This arrangement ensures an automatic gas purge when the cell is opened and also serves as a control of the tightness of the cell at the same time. It also avoids the continual passage of gas through the system, which would place extreme demands on drying the gas.

A few commercial titrators include drying systems to keep the atmospheric moisture outside when renewing the working medium. There are also pump units to change the solvent automatically thus preventing the penetration of humidity.

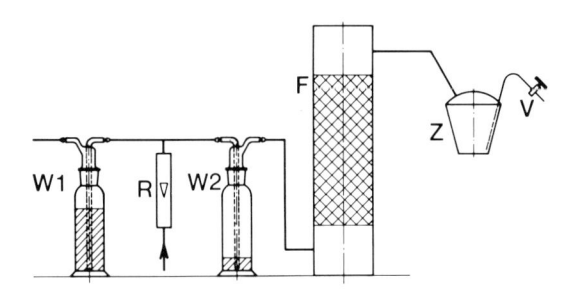

Fig. 6-1 Arrangement to rinse the titration vessel with dried air

A well dried cell consumes no more than 0.02–0.05 ml of KF reagent per hour. This corresponds to approximately 0.1–0.25 mg of water. In practice such values are hardly realized and are not necessary for routine work. The permitted 'drift' depends on the desired accuracy. In case of doubt, the drift can be determined separately by switching the instrument to continuous titration either before or after the water determination in the sample. Calculating the water content of the sample, the drift has to be taken in account. If the drift seems to be too high, any or all of the above methods can be applied as required.

6.5 Indication

The KF titration is self-indicating, but the end point is influenced by the sometimes delayed equilibrium attainment and by moisture from outside (see Section 3.7). Due to the basic yellow colour of the solution, a small excess of iodine can only be detected with difficulty. In practice, therefore, it is always titrated to a "significantly visible" brown colouration, which relies on subjective evaluation. Objective disturbances also influence the identification of the end point. The end-point attainment depends, for example, on the amount of water titrated. With very small amounts of water, the yellow background colour is weak and the equilibrium attainment slow so that the titration is often ended to early. With large amounts of water, the solution becomes intense yellow in colour before the equivalence point is reached, and the equilibrium attainment occurs more rapidly due to the accumulated excess of pyridine and sulphur dioxide. A greater excess of iodine is required here and it tends to be further over-titrated.

Attempts to improve visual indication by the addition of indication aids remained unsuccessful. Starch does not react in non-aqueous solutions. The addition of complementary dyestuffs like methylene blue is unsatisfactory, as the yellow colouration, which is to be compensated for prior to the equivalence point, depends on the working parameters.

Indication can be improved by accelerating the equilibrium attainment of the KF reaction. Increasing the pH 6–7 proves to be very efficient as the concentration of free SO_2 is very low at this pH and hence the yellow colouration also. This way, the KF reaction is also accelerated and the end point is adjusted immediately. Pyridine-free reagents are therefore to be preferentially used. The most suitable are two-component reagents containing sodium acetate as the base as they have the weakest background colour. At low water contents, the solvent component is virtually colourless just before the equivalence point and the change to brown is very distinct. Also, other two-component reagents based on diethanolamine and one-component reagents with imidazole are better suited than the conventional pyridine reagents. A further improvement can be achieved by stabilizing the pH to 7 by the addition of buffer solutions.

Similary, instrumental indication is not without its problems. It was shown during the attempt of the bipotentiometric titration (refer to Section 5.1.2) that the equivalence point can not be directly ascertained here either, again because of a delayed attainment of equilibrium. Using instrumental indication, the titration is also carried out to a slight excess of iodine, approximately in the EP range, shown in the titration curves in Fig. 5–4. Hence, the cut-off point is objectified and the subjective error of the evaluation is eliminated. The error, caused by the dragging attainment of the equilibrium, still remains even when using instrumental indication. The instrument can cut-off here too early, especially when the amount of water is small. The risk of over-titrating at higher water amounts is eliminated. A delayed cut-off reduces the risks. The delay time is usually 10 to 20 s.

Unfortunately, biamperometric/bipotentiometric indication is not definite in each case, as the titration is not carried out to a predetermined excess of iodine but rather to a predetermined signal, a current or a potential. This signal is not produced only by the iodine concentration, but it is influenced by the solvent or by the content of other substances like salts. Thus, for titrations in solvents of low conductivity (2-propanol, tert-butanol), the normal cut-off current of 20 µA for biamperometric titrations for example can sometimes not be achieved, even if a corresponding excess of iodine is present. The instrument continues to titrate to a higher iodine excess and the result is in error. Similar errors occur when hydrocarbons or other non-alcoholic solvents are used in high concentrations. The errors are somewhat lower for bipotentiometric titrations.

The errors thus caused are not usually too serious and can normally be prevented. They can be detected visually by the colour intensity at the equivalence point. They can often be corrected by appropriately adjusting the instrument. Otherwise there is always the possibility of altering the solution composition; usually the methanol content has to be increased. There is also the possibility of adding a small amount of water (0.05%) to the solvent used. Thus a larger amount of KF reagent is used for the pre-titration and the conductive salts thereby liberated improve the indication.

These indication problems are restricted to the use of one-component reagents. Using two-component reagents, a suitable solvent and a sufficient concentration of a conductive salt are introduced by the solvent component so that the indication is distinct and sharp. The rapid reaction of the two-component reagents (see Section 5.1.2) can sometimes cause an over-titration, preferably at the pre-titration of the solvent. An automatic titrator will normally reduce the delivery rate as the end point is neared and this is induced by a slight change of the indicating signal. By the rapid attainment of equilibrium, a pre-signal is not apparent for a two-component reagent, the end point is arrived at "too suddenly" and the instrument cuts off too late. In order to prevent these errors, the delivery rate from the burette is reduced accordingly. Otherwise a small addition of water to the solvent component or a dilution of the solvent with methanol can help.

Irregular end-point indications can sometimes also be observed which do not depend on the water content or on an excess of iodine. N,N-diethyl-p-phenylenediamine for example, induces an immediate end-point indication and an amount of water added can not be recovered. The electrodes are apparently depolarized by this substance.

6.6 Sample Handling

It is an old rule that an analysis can not be better than the sample taken. As every analytical laboratory has its own diverse experience in the handling of samples, general recommendations must not be mentioned. Due to the ubiquitous presence of water, the handling of KF samples requires several special procedures. Thus a number of possible measures will be covered in this section, and everyone can decide for himself which of the possibilities apply to his particular case.

6.6.1 Taking Samples and Sample Storage

It goes without saying that a sample taken must be representative of the average of the product. The sample must also contain the average amount of water and this is sometimes much more difficult to realize. Many substances are hygroscopic and absorb water on their surfaces. The upper layers of a hygroscopic salt for example, can therefore contain a higher percentage of water. It is therefore not an average sample for a water determination, even though it may be representative for all other inspection criteria including a determination of its active constituents. ASTM D 1348 for example, specifically points out that the moisture in cellulose is not uniformly distributed [7906]. The same is true of liquids, especially if they are viscous and denser than water. Here, the upper layer has a higher than average water content. The same considerations apply to the evaporation of water, i.e. to slightly efflorescent hydrates for example or to hydrophobic solvents like benzene.

The pre-treatment of sample bottles also requires particular attention. Plastics are not completely impervious to water vapour and they absorb water on their surfaces, which can equally be desorbed again. For the storage of samples which contain water in the ppm range, glass bottles which have been pre-dried in a dessicator are to be preferred. In extreme cases it can even be necessary to dry the bottles in a drying cabinet and purge them with nitrogen prior to cooling [6301]. Moisture could otherwise recondense on the walls. For storing liquid samples, it proves very practical to rinse the bottles 2 or 3 times with the liquid under investigation before taking the sample. The storage of samples is equally not completely uncritical. If a filled sample bottle is left to stand near a cold wall, the moisture will tend to accumulate on the colder side if the sample is of a solid nature. Changes can even take place in liquids, especially if they are hydrophobic solvents. If a bottle is filled with a saturated xylene sample at 40 °C, the water will separate out upon cooling and the subsequent analysis will result in erratic and incorrect values. In this case, either an immediate analysis, or storage at 40 °C, or the addition of alcohol to act as a supplementary solvent is to be recommended.

6.6.2 Administration of Samples

For a water determination, the sample or a representative portion thereof must be introduced into the titration vessel without its water content being changed. As most products are manufactured and used under normal conditions, the sample can also be handled under normal conditions. Manipulation times should be kept as short as possible. If possible the KF laboratory should be equipped with air conditioning and the KF equipment should not stand near to either moisture sources or drying facilities. Administering samples under extreme conditions like in a dry box for example, should remain the exception and only justified in cases comparable to those like weighing out frozen samples. These samples must otherwise be allowed to reach room temperature in sample vessels sealed in order to prevent the condensation of moisture during weighing.

6.6.3 Liquids

Because of their nature, liquids are easy to administer, preferably by the use of pipettes. It goes without saying that the equipment to be used must be absolutely dry. After cleaning, logically it should be rinsed with methanol and placed in a drying rack for several hours to dry completely. If the complete dryness is doubted, the first sample should be used to rinse the pipette and then discarded. The second sample taken should then be used for the analysis. Reusing pipettes can also prove critical. A pipette containing residues of a hygroscopic liquid will absorb moisture from the atmosphere. It should be considered a matter of course, that pipettes be not filled by suction of the mouth but rather always using a suitable pipette aid.

A further administration apparatus is the injection syringe. By using such a syringe, a sample can be administered into a closed cell through a septum. The cell must not be opened and the atmospheric moisture is completely excluded. This technique is applied preferentially to coulometric water determinations as extremely high demands are placed on the dryness of the system. If glass syringes are used, they should be carefully cleaned and dried at elevated temperatures following each use. It is also to be recommended to cool the syringes in dessicators. Disposable plastic syringes should be rinsed out 2 or 3 times with the sample solution before being used.

6.6.4 Solids

Weighing out solids for KF titrations requires the same basic rules as normal quantitative analysis. Particular attention is called for titrating adherent moisture. Here, the moisture contents to be determined are often less than 0.1 % and these can rapidly change upon contact with atmospheric air. In such cases, sealed weighing vessels are to be preferred.

Putting the sample into the titration vessel can sometimes be problematical, too. Certain titration vessels often have a fixture for this, for example a delivery port for solids including a spoon for direct quantitative weighing of the sample. In any case the titration cell has to be opened, which can lead to the infiltration of an unknown quantity of atmospheric moisture. For this reason a rinsing with dry air is to be recommended (see Section 6.4). Sometimes, however, the sample can be dissolved in a measuring flask and administered in liquid form. This mode of practice also entails the risk of a blank determination and the possible hygroscopic nature of the solvent. It should therefore only be used as a last resort, for example if a large sample is necessary because of homogenity reasons and the sample has a water content which is too high.

To determine the water quantitively, it is normally necessary to dissolve the sample completely in the solvent or in the working medium. Otherwise the water must diffuse out of the undissolved substance and this results in dragging titrations, fading end points and exceedingly low results. In such cases the water titration should not be started before the substance is completely dissolved. In certain cases, the water is completely and rapidly released without the substance having dissolved. This is especially true for a few hydrates of inorganic salts, for example sodium sulphate-10-hydrate.

Many naturally occurring products are insoluble, and the extraction of water requires more time. In order to eliminate additional sources of error, it is desirable to extract the substances directly in the titration cell. The titration at elevated temperature as described in Section 3.5 is a method to be recommended. Under such conditions, the titration is usually completed within about 15 min.

If the extraction takes longer than 15–30 min, it should be carried out outside of the titration vessel. In the simplest case, the sample can be extracted in a measuring flask using a suitable solvent which possibly has been dehydrated by a pre-titration. Extraction times can be as long as several days without any interference due to foreign moisture taking place. Extraction times can be shortened by using appropriate solvents at elevated temperatures.

If deemed necessary, the sample can also be extracted by refluxing in a suitably moisture-sealed apparatus. This method of extraction should only then be applied when all other methods have proved unsuccessful. Moisture can cause severe disturbances here, and the subsequent technique employed should always be checked for reliability by a blank determination. Extraction techniques are above all applicable to investigation of foodstuffs and thus are more fully described in Section 9.1.

6.6.5 Gases

There are a number of substances like air, nitrogen, oxygen, low hydrocarbons or halogenated hydrocarbons which are gaseous at room temperature. Water can also be determined in these substances by passing the gas to be determined through a absorption liquid, like methanol for example [6101, 7907]. The water is thereby absorbed by the liquid and can thus be determined by a Karl Fischer titration.

For the investigation of gases, a great deal of special apparatus has been described in the last decades (refer to Section 10.1) which enable the absorption of moisture and its titration in a normal vessel. Today, one usually resorts to commercially available equipment. These cells have either been specifically constructed for the study of gases or need only a suitable gas inlet tube.

Taking or storing a gaseous sample can present serious problems, as instruments with sealing liquids can not be used. To add to these difficulties, many gases contain only ppm amounts of water, which means that large volumes of gas are required. It therefore proves practical to introduce the sample directly at its source into a suitable absorption vessel, transferring thus the water into a liquid phase which is more easy to handle. As moisture can be easily adsorbed as well as desorbed from the walls of tubes, all connections should be thoroughly purged with the gas so that a stable equilibrium is attained prior to the determination.

The simplest way of absorbing the moisture is to use a gas washing bottle which contains a suitable adsorption liquid. At the sample source, an appropriate volume of gas is passed through the wash bottle, after which the adsorption liquid or an aliquote portion thereof is transferred into a KF titration cell and titrated.

It is more elegant and reliable to use the titration cell of the KF titrator to adsorb the moisture. The cell is filled as usual with a suitable solvent and dehydrated by pre-titration. The sealed cell is then dismantled from the instrument and brought to the source of the sample. There, a known amount of gas is passed through the cell. Then the cell is connected to the instrument and the titration is carried out. This method is particularly practical for determining relatively small amounts of water as it eliminates secondary sources of error due to handling the adsorption liquid.

Methanol is often used as the absorption medium as it is generally the common KF solvent. If several litres of a gas are passed through the absorption medium, losses due to evaporation should be taken into account. Therefore, for dry gases which require particularly high sample volumes, solvents with high boiling points should be preferentially used, ethylene glycol for example, or mixtures of ethylene glycol and pyridine or ethylene glycol and the solvent component of a two-component reagent.

A diluted KF solution is often recommended as the titration agent because the amount of water is relatively low. The strength of the titrating agent depends primarily on the expertise of the working technique employed. The basic rule of the KF titration also applies here, i.e. that with increasing dilution of the titrant, the susceptibility to errors due to unwanted moisture also increases.

The coulometric determination of water is to be particularly recommended for the analysis of gases as it is a micro-method and thus only requires small samples. It is preferred to pass the sample directly into the KF cell, which sometimes has to be dismantled from the instrument. As the gas to be investigated only passes through the cell and does not otherwise contaminate the electrolyte, it is thus possible to carry out many determinations in succession. This is a particular advantage of the coulometric determination of water in gases.

This technique can be varied for the investigation of liquified gases. These substances are stored in pressurized vessels in liquid form and samples can be stored and transported under such conditions. When introduced into the titration cell, the sample must be taken from the liquid side of the stock vessel. If a gaseous sample is taken, i.e. the sample is allowed to boil in the stock bottle, the gas phase as a rule will contain less water than the liquid phase. Liquid gases can also be condensed into a deep-cooled titration cell. The condensation of methylene chloride for example (boiling point $-23\,°C$) into pre-cooled methanol is possible, the subsequent water titration is carried out at $-30\,°C$. For such titration methods, two-component reagents are to be preferred as they still ensure a sufficiently adequate titration rate even at these low temperatures.

Further details concerning the analysis of certain technical gases are to be found in Section 10.1.

6.7 Safety Precautions

Safety precautions in laboratories are being taken more and more into consideration mainly because it has since been recognized that even the smallest health hazard should be prevented if possible. Adequate safety observations, however, require to know the products and the potential dangers. There appears to exist a lack of information here, mainly because the sources of information are not known.

A number of product informations can be obtained easily. In the states of the European Economic Community, dangerous substances must be marked in accordance with the provisions now laid down in the Chemical Law (of 19th September 1980) by means of hazard symbols, special risk phrases (R phrases) and safety phrases (S phrases). These informations have to be printed on the label of each container, thus informing the user.

6.7 Safety Precautions

The "Registry of Toxic Effects of Chemical Substances" – a monograph – is published by the National Institute for Occupational Safety and Health (NIOSH). "It is intended to provide basic information on the known toxic and biological effects of chemical substances for the use of employers, employees, ... and, in general, anyone concerned with the proper and safe handling of chemicals. In turn, this information may contribute to a better understanding of potential occupational hazards by everyone involved and ultimately may help to bring about a more healthful workplace environment." (Foreword to the 1979 edition). The RTECS describes toxicity data, mutation data, skin and eye irritation data and threshold limit values (TLV) for substances in working rooms. It is updated continuously and published annually.

The "Toxic and Hazardous Industrial Chemicals Safety Manual" is published in Japan by the International Technical Information Institute. This manual is a result of the study of literature and expertise in Japan and the world. It contains information on properties, hazardous potentials, toxicity, handling, storage and waste treatment of about 700 industrial chemicals.

In Germany "Gefährliche chemische Stoffe" are described in Appendix 3 of the "Unfallverhütungsvorschriften" (UVV) of the "Berufsgenossenschaft der chemischen Industrie". This appendix not only describes the important physical properties but also effects and the necessary preventative measures of the more common dangerous chemicals. Appendix 4 of the UVV, "MAK-Werte" (Maximum workplace exposure levels), is revised and published annually. The "Merkblätter Gefährliche Arbeitsstoffe" from Kühn-Birett (published by Verlag Moderne Chemie) are much more detailed and describe physiological effects as well as giving technical details; and refer to toxicological data as well as giving information on possible medical treatment.

A lot of product information available has been compiled in manufacturing or processing plants and is well known to the cognizant laboratories. For several substances which are predominantly used for KF titrations, the data shall be reproduced such as is known to the author and appears to be of importance.

Methanol, by far the working medium most used for KF titrations, is primarily to be regarded as a fire hazard as its flash point is 11 °C and its explosion limits are 5.5 and 31 % by volume. Inhalation can cause dizziness, headache, weakness, vomiting and other disorders. The well-known blindness due to methanol occurs less often than assumed. R-phrases: Highly flammable. Toxic by inhalation, in contact with skin and if swallowed. The acute toxicity is relatively low (MAK value: 200 ppm, TWA: 200 ppm).

Pyridine, the classic base in KF reagents, is classified as being hazardous to health and inflammable. Swallowing induces coughing, dizziness and headache, larger amounts can lead to vomiting, circulation disorders and dizziness. The same symptoms can be observed after inhalation or upon skin contact. R-phrases: Flammable. Harmfull by inhalation, in contact with skin and if swallowed. MAK value: 5 ppm, TWA: 5 ppm.

2-Methoxyethanol, the standard solvent for one-component reagents, is injurious to the central nervous system. In spite of its low vapour pressure, toxically chronic amounts can be absorbed by inhalation. R-phrases: Flammable. Irritating to eyes. Recent investigations require increased precaution. On experimental animals 2-methoxyethanol has shown haematological effects characterized by reduction in both red and white blood cells, bone marrow and lymphe node depressions, testicular atrophy, foetotoxicity and teratogenicity. Inhalation and contact with the skin should be prevented and it should only be used in sealed vessels if possible and waste be properly disposed of. The MAK value has been reset to 5 ppm.

Chloroform, the solvent agent for fats, has the property of causing irritation of the skin and conjunctiva. Its narcotic effect is well known. R-phrases: Harmfull by inhalation. MAK value: 10 ppm (approx. 50 mg/l), TWA: 50 ppm. Chloroform is also one of those substances believed to be potentially carcinogenic.

Formamide is an excellent dissolver for many substances and it is used as an additive in KF titrations. It has a very low vapour pressure. R-phrases: Irritating to eyes, respiratory system and skin. Possible risks of irreversible effects. The MAK value is given as 20 ppm (approx. 37 mg/l), TLV: 20 ppm. Its physiological effects remain unclear. It is sometimes classed as being carcinogenic. Based on experimental results on animals, swallowing, skin contact or inhalation can cause teratogenic effects and its use requires particular care: avoid all skin contact and inhalation, use only in sealed containers and ensure controlled disposal of waste KF solutions.

As the hazardous substances can not always be replaced by other substances, the application of the KF titration technique is also a question of safety. Due to the necessary exclusion of moisture, the actual titration is carried out in a closed container so that the safety requirements are thus automatically met. In order to administer the working medium and solvent, pipettes, flask dispensers or plunger burettes which permit a hermetically sealed dosage should be preferentially used. Attention to KF waste is also called for. Immediate disposal into running water down the drain is of course to be recommended if this is technically feasible (wastewater considerations). The other alternative is to collect the waste solutions in a sealed waste container whereby the compatibility of the individual waste constituents also has to be considered.

7 Organic Compounds

The main field of application of the KF titration is in the investigation of organic compounds and their commercially viable technical products. These substances are usually sufficiently soluble in methanol although in certain cases the working medium has to be adjusted according to the solubility requirements. Certain functional groups (like mercaptans and aldehydes for example), can cause side reactions which can make the water determination more difficult. The procedure to determine the water content and the disturbances which can take place are mentioned in each section covering these substances. Generally speaking, interferences are more seldom than would be expected from the number of functional groups considered.

7.1 Hydrocarbons

Saturated liquid hydrocarbons can be titrated by the application of standard methods without any difficulty. An alcoholic working medium is a prerequisite for a stoichiometric reaction and a reliable end-point indication. Earlier studies which recommend a direct titration (without the addition of alcohol) should be treated with scepsis. They only give a definite end point when the water content is high. Then a bigger amount of KF reagent is required, thus introducing a sufficient excess of alcohol to ensure the end-point adjustment. We could not confirm the observations sometimes made, that exact results are only obtained by employing a back-titration method. Presumably the higher alcohol content, added this way, had been the reason for the better results.

The limited solubility of higher hydrocarbons in methanol can lead to problems as they form two separate layers in the titration vessel. Although the total amount of water contained will diffuse into the methanolic phase during the titration, this can take time especially if stirring is inadequate. Indication problems can also occur when the indicator electrodes come into contact with the hydrocarbon phase. Additives which improve the solubility and enable the titration in a homogeneous phase are therefore practical. 1-Propanol or a methanol/chloroform mixture (1 + 2 parts by volume) can be used instead of pure methanol if the titration is carried out using a one-component reagent. If a two-component reagent is being used, the solvent component can be partially replaced by 1-propanol or chloroform.

The lower hydrocarbons are gaseous. The techniques necessary for these are described in Sections 6.6.5 and 10.1.

Unsatured hydrocarbons can normally be titrated in the same way as saturated compounds. Side reactions due to addition to double bonds only occur in exceptional cases, for example with allocymol or with norbonadiene-(2,5) [Hyd. p. 26]. Here, an end point is either not obtained or is not stable, or the amounts of water which are found are too high. These effects can be avoided or at least reduced by using long-chain alcohols instead of methanol as the working medium and a titrant based on 2-methoxyethanol.

Hydrocarbons mostly contain very little water, usually less than 500 ppm. The KF titration should therefore be carried out as a micro-determination employing titrants with a WE of 1 mg/ml. In this case the course of the titration can be very dragging, however, and Muroi and Ono [7101] recommend adding pyridine and sulphur dioxide to the working medium. Two-component reagents are to be preferred as they show a quick and stable attainment of the end point. Also, at such a low titre, non-hygroscopic titrants are to be recommended as they are insensitive to atmospheric moisture.

Coulometry is without doubt the ideal method of determining water in hydrocarbons as it already is a micro-method. The other requirements – liquid samples and goot compatibility – are also met sufficiently. The KF reagent should be so chosen that it has a good solubility for hydrocarbons. It should contain a mixture of methanol and chloroform or a long-chain alcohol.

7.2 Halogenated Hydrocarbons

Just like hydrocarbons, halogenated hydrocarbons can be analysed without any difficulty. The water determination is hardly affected provided sufficient alcohol is present. The standard technique is the preferred working method. Long-chain halogenated hydrocarbons can cause the same problems as with their homologous hydrocarbons for reasons of solubility, and it is therefore practical to modify the working medium here as well. 1-Propanol is equally to be recommended as a solvent as are mixtures of methanol and chloroform. When two-component reagents are used, the solvent component should be diluted with chloroform or 1-propanol.

Several halogenated hydrocarbons like chloromethane and difluorochloromethane are gaseous at room temperature. These substances can be analysed according to the general working procedures for gases, which are to be found in Sections 6.6.5 and 10.1. A titration in methanol cooled to $-30\,°C$ is also possible. Methanol is sufficiently cooled and pre-dehydrated. The gas to be investigated is then passed through the cell where it condenses. As the following KF titration takes longer at such low temperatures, the "quick" reagents prove more practical, pyridine-free two-component reagents in particular. The KF titration can also be carried out in pressurized titration cells. This working technique has been applied to investigation of fluorinated hydrocarbons and aerosols [Aqu. p. 354].

Chlorinated hydrocarbons sometimes contain free chlorine. This affects the water determination as the chlorine oxidizes the iodide in the reagent to iodine and consumes water. This can result in a too low or even a "negative" water content. According to Muroi, Ogawa and Ishii [6701], this error can be avoided by pre-treating the sample with a solution of sulphur dioxide and pyridine in methanol. The chlorine is reduced by the sulphur dioxide and the water can then be titrated. The sulphur diox-

7.3 Alcohols and Phenols

ide/pyridine solution must be free of iodide, however, and therefore can not be dehydrated by means of a pre-titration. The blank value must be determined separately and allowed for in the calculations. Effects due to chlorine can also be eliminated by the addition of an unsaturated hydrocarbon like tetradecene to the sample. The chlorine is added to the double bond in a few minutes and the water can then be titrated.

Coulometry can also be successfully applied to halogenated hydrocarbons as these compounds also contain only small amounts of water. It also appears to be recommendable for gaseous compounds.

7.3 Alcohols and Phenols

The determination of water in alcohols is possible without any disturbing effects as alcohols constitute one of the components of the KF solution. A number of alcohols can be used to replace methanol, as in the water determination of ketones for example. A direct titration of the water content is therefore sometimes possible though the standard technique should be preferentially applied.

Long-chain alcohols tend to be more soluble than their hydrocarbon counterparts and additives like 1-propanol, which improve the solubility, are only necessary with very long carbon chains (stearyl alcohol). Highly viscous alcohols (glycerol) sometimes require a thinning agent in order to improve the sample administration. Higher alcohols (sorbit) can present solubility problems (see Section 9.3 "Carbohydrates"). Phenols are solids and can be titrated by using standard techniques. Oxidation by iodine does not appear to occur, so that water can even be determined in hydroquinone.

In alkali alcoholates, the water is present as the alkali hydroxide. It can be titrated if the substance is treated with acid. The alcoholate is thereby transformed to its alcohol according to Eq. (7.1) and the alkali hydroxide to water according to Eq. (7.2). In order to carry out the titration, an excess of benzoic acid or propionic acid is added to the working medium beforehand and dehydrated. The alkali alcoholate is then added and titrated. The extremely hygroscopic nature of the substance calls for particular care when handling the sample.

$$CH_3ONa + HX \rightarrow CH_3OH + NaX \qquad (7.1)$$

$$NaOH + HX \rightarrow H_2O + NaX \qquad (7.2)$$

The coulometric determination of water can equally be undertaken. But many alcohols contain larger amounts of water (more than 1%) so that either the determination takes longer or micro-amounts of samples have to be administered. If particular accuracy is required, a volumetric determination should be preferred. Coulometry can not be recommended for phenol without any restriction. At higher concentrations, we found an irreversible poisoning of the anode which then had to be cleaned using sulphuric/chromic acid. Diphenols (hydroquinone) can not be analysed, as they are anodically oxidized and do not give an end point.

7.4 Ethers

Ethers and phenol ethers act like hydrocarbons and can be analysed the same way. Long-chain ethers can require a solubilizer. Ethylene oxide, propylene oxide and butylene oxide are also analysed in the same way. Dimethyl ether (boiling point −23 °C) is difficult to titrate. The water reacts only very slowly with the conventional KF reagents at low temperature (−40 °C). It can be recommended to add dimethyl ether to pre-dehydrated methanol, to evaporate the ether and to titrate the remaining water at room temperature [Aqu. p. 319].

Vinyl ethers react with KF reagents which contain methanol. Barnes and Pawlak [5902] assumed a reaction as in Eq. (7.3). They replaced the methanol in the system by using a KF reagent based on 2-methoxyethanol and pyridine/acetic acid (5+2) as the working medium.

$$ROCH = CH_2 + I_2 + CH_3OH \rightarrow ROCH(OCH_3)CH_2I + HI \tag{7.3}$$

Polyether has been analysed by Andrysiak and Andrysiak [7201].

7.5 Aldehydes and Ketones

These two classes of compounds are the problem groups of the Karl Fischer titration. They react with the methanol in the KF reagent or working medium according to Eqs. (7.4) and (7.5) forming acetals and ketals. The water formed in this process is titrated simultaneously, thus producing vanishing end points and erroneous high results.

$$\underset{R}{\overset{H}{>}}C=O + \underset{CH_3OH}{CH_3OH} \rightarrow \underset{R}{\overset{H}{>}}C\underset{OCH_3}{\overset{OCH_3}{<}} + H_2O \tag{7.4}$$

$$\underset{R}{\overset{R}{>}}C=O + \underset{CH_3OH}{CH_3OH} \rightarrow \underset{R}{\overset{R}{>}}C\underset{OCH_3}{\overset{OCH_3}{<}} + H_2O \tag{7.5}$$

Aldehydes react at very differing rates. Acetaldehyde reacts most quickly and causes the most problems during water determination. The reactivity decreases with increasing chain length. Aromatic aldehydes react more slowly than aliphatic aldehydes. Formaldehyde and chloral are exceptions to this rule as they do not form acetals and can be titrated without any difficulty.

Ketones are less reactive than aldehydes but still show the same gradation within the ketone group. Acetone or cyclohexanone react most quickly, and the reactivity decreases with increasing chain length. Diisopropyl ketone, benzophenone, desoxybenzoin, benzil, benzoin, camphor, dibenzalacetone and dichlorotetrafluoroacetone are all considered stable [Aqu. p. 381].

The formation of acetals and ketals is also influenced by the reactivity of the alcohol present. Methanol reacts the most quickly and the rates of reaction decrease with increasing chain length. The KF reagent used also appears to have an influence. Two-

7.5 Aldehydes and Ketones

component reagents show a higher reactivity than one-component reagents, possibly caused by the higher concentration of alkyl sulphite.

The investigation of aldehydes is influenced by a second side reaction, the well-known bisulphite addition according to Eq. (7.6). In this reaction, water is being consumed and the water contents found are too low. Ketones react similar to a certain degree, but much more slowly.

$$\begin{array}{c} H \\ R \end{array} \!\!\!> \!C = O + SO_2 + H_2O + NR' \rightarrow \begin{array}{c} H \\ R \end{array} \!\!\!> \!C \!\!\!< \begin{array}{c} SO_3HNR' \\ OH \end{array} \tag{7.6}$$

A number of attempts have been undertaken to modify the working conditions or the reaction medium so as to enable a determination of water in aldehydes and ketones.

Bryant, Mitchell and Smith [4001] recommended the addition of hydrocyanic acid to aldehydes and ketones according to Eq. (7.7). The cyanhydrine thus formed does not affect the KF titration. This technique does not appear to be widely practised.

$$\begin{array}{c} H \\ R \end{array} \!\!\!> \!C = O + HCN \rightarrow \begin{array}{c} H \\ R \end{array} \!\!\!> \!C \!\!\!< \begin{array}{c} CN \\ OH \end{array} \tag{7.7}$$

Wernimont and Hopkinson [4301] applied their dead-stop technique to the determination of water in acetone. They used a reagent low in methanol (see Table 4–2) for this and the back-titration method. Reliable results were obtained with pure acetone. The method proved unsatisfactory for acetone/alcohol mixtures, containing more than 15% alcohol.

Eberius [Eb. p. 113] recommends the use of pyridine as the working medium and a "normal" (methanolic) KF reagent.

Fischer and Schiene [6401] replaced the methanol completely. They used pyridine as the working medium and the KF reagent of Peters and Jungnickel [5501] based on 2-methoxyethanol. The determination of water in acetone and in methyl ethyl ketone was "well reproducible".

Applying this procedure to aldehydes, Beyer and Varga [6601] did not obtain stable end points (formation of acetals) and the water content found was too low (bisulphite addition). They recommended a titration with a methanolic KF reagent at $-10\,°C$. The acetal formation and bisulphite addition are thereby suppressed.

Muroi, Ogawa and Ishii [6502] used a mixture of pyridine and glycol (3 + 1) as the working medium and a KF reagent based on tert-butanol/chloroform (Mutsubishi). For the investigation of acetaldehyde, the sample is added to the working medium and evaporated by a stream of dry nitrogen. The water remains in the working medium and is titrated in the presence of any residual acetaldehyde. The water content in polymeric formaldehyde was also studied using the same reagents [6302, 6303].

Klimova, Sherman and L'vov [6703, 6704] work in a completely alcohol-free system. They use dimethylformamide as the working medium and also a KF reagent based on dimethylformamide (see Section 4.1).

Muroi an Fujino [8304] recommended a mixture of chloroform, propylene carbonate, pyridine and sulphur dioxide as a working medium and an alcohol-free titrant to investigate ketones.

There is no working technique universally applicable to ketones and aldehydes, and in many laboratories specific variations have been developed to solve specific problems. Most of these modifications consist of working in non-methanolic media. Pyridine, mixtures of pyridine and ethylene glycol or 2-methoxyethanol are preferred as the working medium. Commercially available reagent solutions based on 2-methoxyethanol or high pyridine formulations as indicated in Table 4–2 serve as the titrants.

The KF reaction is quicker than the side reactions. This is often used to determine water in carbonyl compounds. One titrates very quickly thus "overtaking" the side reactions. In order to finish the titration quickly, it is greatly over-titrated. At the same time, the persisting time at the end point is drastically reduced, to 3 or 5 s for example. This technique can be applied to both visual as well as instrumental indication. Titrating at 0 °C also makes use of the differing reaction rates. The side reactions are sufficiently delayed at this temperature to enable the KF titration to be carried out more easily. This technique can be recommended for aldehydes.

The graphical evaluation of the course of a titration, as shown in Figs. 3–1 and 3–2, also offers the possibility of differentiating between a KF reaction and the formation of ketals [Hyd. p. 25]. First, the free available water reacts very rapidly, followed by a small, continually consumption of reagent, caused by the formation of ketals or acetals. This can be seen graphically from the ascending branch of the curve. Extrapolation of the curve to the time zero cancels out the ketal formation. Dirscherl [8208] applied the graphic evaluation to a micro-method. Ketones are titrated at room temperatures, aldehydes preferably at 0 °C.

Comprising all the methods, the determination of water in aldehydes and ketones can only be considered as being satisfactory, because the otherwise usual accuracy is achieved only occasionally. The applied working technique should sometimes be evaluated critically. If recovery rates of water are to be determined, several different amounts of water should be added in order to be certain that the formation of ketals and a bisulphite addition do not compensate for each other, thus creating the impression of a reliability which does not exist.

In coulometry, aldehydes and ketones present the same problems and methanol has to be replaced here as well. Petrov, Galitsyn and Kasperovich [8007] use 2-ethoxyethanol as the solvent in their coulometric reagent. Mitsubishi has applied for patents for their formulation using alkyl carbonates [8005] and market such a reagent containing 14% propylene carbonate, 62% chloroform, 21% pyridine and sulphur dioxide [8304].

In 1984 Riedel-de Haën introduced a new Hydranal reagent, especially designed for volumetric titration of water in carbonyl compounds. Ketones as well as aldehydes can be titrated at room temperature. A new alcohol used as the working medium suppresses the formation of ketals and acetals and ensures stable end points.

7.6 Acids, Esters and Salts

The majority of acids allow a trouble-free water determination. They react weakly acidic, influence the pH of the KF system only slightly and do not otherwise affect the KF reaction. The water determination of many acids has been investigated [Aqu. p. 308–309, 8201, 8301].

7.6 Acids, Esters and Salts

Several possible interferences must be mentioned, the first being that of esterification. During titration this is indicated by a fading end point and by results which are sometimes too high. The tendency to esterify is strongest with formic acid. It reacts so rapidly with methanol that a water determination is not possible using two-component reagents which are particularly reactive. One-component reagents only enable a titration when the amount of formic acid is limited, for example to 1 ml formic acid in 20 ml methanol [8201]. If methanol is replaced by a less reactive alcohol like 2-methoxyethanol or 1-propanol for example, larger amounts of formic acids can be titrated.

Acetic acid also esterifies, though more slowly. Esterification is only significant at higher concentrations of approximately 5 ml per 20 ml methanol and above. An inactive alcohol is to be preferred here as well. The tendency to esterify decreases with increasing chain length. Propionic acid does not esterify during normal volumetric titrations. With the more sensitive coulometric determination, the esterification is clearly evident, and the amounts to be applied are thus limited [8301]. Aromatic acids do not esterify. We also found a tendency of bromoacetic acid, dichloroacetic acid, and anhydrous oxalic acid to esterify [8201] so that the amount to be titrated must be restricted to 5 g or even 2 g per 20 ml methanol.

The esterification can also be delayed by neutralizing the acid with pyridine or another base [Eb. p. 79, Hyd. p. 30–32]. This method is less effective than the replacement of methanol by an inactive alcohol.

Stronger acids like tartaric acid, citric acid and trichloroacetic acid, when used in high concentrations, can shift the pH of the KF reaction far enough into the acidic region. The KF reaction is delayed and a titration end point is either not reached at all or only very slowly. These acids therefore have to be neutralized prior to a water determination. Only amines can be used to neutralize the acids, as they do not produce water as a result of the neutralization reaction. Amines are never completely free of water and they must be dehydrated by pre-titration together with the solvent. Thus only medium-strength bases (pyridine, imidazole) are suitable. Strong bases would alkalize the solvent thus inhibiting the pre-titration. Buffer solutions can also be recommended for this purpose as they ensure a stablized pH during the pre-titration and the following water determination. Therefore, the KF reaction takes place stoichiometrically and swiftly in the proper pH range.

In order to carry out such a titration, the standard technique is modified. An amount of amine, sufficient to neutralize the sample of the acid, is added to the working medium and dried by pre-titration. Then the sample is added and titrated in the usual manner.

Interferences due to other functional groups of acids can occur as well. Keto acids act like ketones and can form ketals. They are to be treated like ketones. With aromatic amino acids, an N-methylation (as with aniline) can not be completely ruled out, and this can also lead to a fading end point [8201]. Ascorbic acid is oxidized by iodine so that a direct water determination is not possible. Using his two-component technique, Johansson [5601] recommended to oxidize the ascorbic acid with the iodine solution first. Then the sulphur dioxide-pyridine component is added, and the water content is titrated.

Esters can be investigated without any difficulty if the standard technique is employed. They are similar to hydrocarbons in that long-chain aliphatic esters can require the addition of a solubilizing agent. Chloroform is a proven additive for fatty acids. In spite of the presence of a double bond, acrylic acid esters can also be investigated without any problems. Esters of keto acids behave like ketones and have to be treated accordingly. The amino acids are practically insoluble in methanol and other organic solvents. For this reason, only the adherent water can be determined.

The same basic rules apply to the salts of carboxylic acids as well as for inorganic salts, as described in Section 8. For soluble salts, the determination of water is without any problems, as they react neutrally and do not alter the pH of the KF system. Sodium tartrate dihydrate and sodium acetate trihydrate are used to standardize the titre of KF solutions. Insoluble salt can cause problems. Adherent moisture is easily determined, but inherent moisture can not usually be titrated. Several hydrates release their water of crystallization easily and can be titrated in suspension.

7.7 Nitrogen-Containing Compounds

7.7.1 Amines and N-Heterocycles

Both these groups of nitrogen-compounds should be considered together as they behave similar and show sometimes the same disturbing effect as well. They can be devided into three groups according to their behaviour; weak bases, strong bases and those substances causing interferences.

Weak bases with pK_B values greater than 8 allow a water determination without any problems. They behave like hydrocarbons and can be titrated as such. This group consists of a number of heterocycles notably pyridine, picoline, quinoline, imidazole, benzimidazole, indole, benzothiazole and nicotine to mention only a few, and also several substituted aromatic amines like N'N-dimethylaniline and diphenylamine.

Strong basic compounds with pK_B values less than 8 can cause problems if they are added in large amounts. They can shift the pH of the Karl-Fischer system into the alkaline range. Titrating water under these conditions, the end points appear much later, the stoichiometry is changed and the results are erroneously high. When strong bases are used in high concentrations, an end point is not reached at all. It is therefore reasonable to neutralize the base prior to titrating. As the acids necessary for this contain water, it is practical to dehydrate them together with the solvent during pre-titration. They can therefore only be weakly acetic as otherwise they would make a pre-titration impossible. Acetic acid is mostly recommended for this and applied in a slight stoichiometric excess relative to the base. This procedure ensures that the pH value of the KF system remains in the optimum range.

The use of acetic acid is not completely without its problems. Muroi and Ono [7003] recommend controlling the acid-base relation and a titration at a low temperature. Investigating methylamine, Cunningham [Aqu. p. 333] found a similar dependancy on the acetic acid concentration. At low acetic acid concentrations, he found water contents which were too high, reproducible and agreeing results in the intermediate range, and drifting end points at high acid concentrations. The cause of fading end points is apparently the esterification of the acetic acid. It is often overlooked. Therefore, the titrations are carried out in cooled solutions, preferably in titration vessels cooled in ice. Propionic acid is more suitable than acetic acid. Cunningham [Aqu.

7.7 Nitrogen-Containing Compounds

p. 337] recommended the use of salicylic acid, whereby cooling is not required. Today we know that salicylic acid does not esterify during titration, neither does benzoic acid which is the acid we prefer [8201]. By using this acid, the KF titration can be carried out at high temperatures, which is often practical for wax-type, long-chain amines. A particular mole ratio of benzoic acid to amine is not necessary as long as the benzoic acid is present in a stoichiometric excess. An excess of benzoic acid does not cause any other effects.

A "spent" Karl-Fischer solution is also a possible neutralization agent for strong bases as it contains hydroiodic acid, methyl sulphuric acid and an excess of methyl sulphurous acid from the previous titration. All these acids are present as their pyridinium salts. These pyridinium salts are weakly acidic and neutralize strong bases. Therefore, the solution can be applied to neutralize basic substances. In this case the acid must be present in a stoichiometric excess, too.

Johansson [5601] had already pointed out that with two-component reagents, the solvent component has a neutralizing effect, as it contains sulphur dioxide. Up to 0.5 g amine can be titrated in 10 ml solvent component. For larger amounts of amine, which are necessary if the water content is low, a neutralization with acetic acid is to be recommended. 10 ml acetic acid are added to 20 ml solvent component and dehydrated by pre-titration. The same considerations also apply to other two-component reagents. In view of esterification, acetic acid should not be used but rather benzoic acid which also is sufficiently soluble in solvent components (300 g/l = approx. 2.5 mol/l).

Several amines, weak or strong bases, show deviating characteristics. For example, Lima and Spiro [7603] did not find an end point for the determination of water in 1,2-diaminoethane, not even after having neutralized the substance. Only a ten-fold excess of acetic acid could enable a titration. Johansson [5601] reported a slightly fading end point with aniline. We found similar difficulties with m-toluidine, aminophenol, 1,2-diaminopropane and 1,2-phenylenediamine. We presume that methylation reactions with nitrogen are the cause of this deviating behaviour. It is therefore practical to use titrating agents which are free of methanol and to carry out the water determination in a methanol-free working medium like methyl glycol for example.

Furthermore, we have observed several irregular deviations which we would like to mention. Titrating pyrrol we got a stable end point during titration but the water content found was too high. Adding water we got recovery rates of about 130%, and the relative error was greater for smaller amounts of water than for larger amounts. 1-Methylpyrrol and indole showed similar errors though not quite as large.

N'N'-diethyl-p-phenylene diamine induced a very rapid end point which did not relate in any way to the water content. Amounts of water added could not be recovered [Hyd. p. 33].

Long-chain amines dissolve only slowly or not completely in methanol so that an addition of chloroform proves practical. For wax-type amines, higher temperatures are employed as these substances then dissolve very easily. For example, one can work with trichloroethylene/methanol or trichloroethylene/methanol/benzoic acid at 50 °C. Both the pre-dehydration and also the titration are carried out at this temperature.

The same basic rules for volumetric titrations also apply to the coulometric determination of water [8301]. Weak bases can be analysed without any problems, strong bases have to be neutralized. Benzoic acid can also be used for a neutralization here, provided it is soluble in the KF reagent. Acetic acid is unsuitable as it esterifies too rapidly. Basically all those known effects also occur in coulometry and usually to a greater extent. This becomes apparent because of the high sensitivity of coulometry and an accumulation of these effects due to successive determinations, which are common in coulometry.

7.7.2 Amides

The amides of carboxylic acids behave like weakly basic amines and the determination of water appears to be without any problems. A series of amides, anilides and urea derivatives has been analysed [Aqu. p. 336]. Two amides are of particular importance to the determination of water: formamide and dimethylformamide. Both are used as solvents or as additives to the working medium (see Section 6.1).

7.7.3 Nitriles and Cyanohydrins

These compounds react weakly basic as well and a water determination can normally be carried out without too much difficulty. Acetonitrile is used by Miyake and Suto together with ethyl glycol (60 + 2) as the solvent for a coulometric reagent [7705]. Also, the cyanohydrins are stable in Karl-Fischer media. They are formed as a result of treating aldehydes and ketones, which disturb the KF titration, with hydrocyanic acid according to a method of Bryant, Smith and Mitchell [4001].

7.7.4 Hydrazine Derivatives

Hydrazine is oxidized by iodine in neutral solutions and the KF titration can only be carried out under acidic conditions [Hyd. p. 36]. The same applies to substituted hydrazine [Aqu. p. 346]. Acetic acid is too weak to prevent the oxidation of hydrazine completely. Dichloroacetic acid or an exactly calculated amount of sulphuric acid should therefore be used instead.

7.7.5 Other Nitrogen Compounds

The determination of water in other nitrogen-containing compounds is also possible. Azo-, nitro-, and nitroso compounds, oximes, hydroxamic acid, isocyanates, purines and amidines were investigated without any further details being given [Aqu. p. 332].

Alkaloids have been very thoroughly studied [Aqu. p. 344]. The water content of approximately 30 substances were compared with loss-on-drying values. In every case, the KF method proved more reliable.

7.8 Sulphur Compounds

Mercaptans are oxidized by the iodine present in the KF reagent:

$$2\,RSH + I_2 \rightarrow RSSR + 2\,HI. \tag{7.8}$$

7.8 Sulphur Compounds

The water determination is not possible directly in these compounds. Several attempts have been undertaken to combine mercaptans with other products, thus enabling the KF titration. It was attempted to add mercaptan to double bonds in the presence of boron trifluoride [Aqu. p. 362–366].

$$\text{RSH} + \text{HC} = \text{CH} \rightarrow \text{RS}-\overset{\overset{R'}{|}}{\underset{\underset{H}{|}}{C}}-\overset{\overset{R'}{|}}{\underset{\underset{H}{|}}{C}}-\text{H}. \tag{7.9}$$

Francis and Persing [7810] recommend the addition of thiols to acrylonitrile or to N-ethylmaleinimide in the presence of pyridine. The reaction takes place very rapidly at room temperatures. Acrylonitrile gives good results with primary thiols and with mercaptocarboxylic acids.

Sulphides, R—S—R′, and disulphides do not present any problems to the KF titration.

Thiourea shows a fading end point applying larger samples. Substituted thioureas show differing characteristics.

Alkali xanthogenates and dithiocarbamates permit a KF titration even though they are oxidized by iodine in aqueous solution [5102].

Dimethyl sulphoxide can alter the stoichiometry of the KF reaction. According to Bykowa et al. [7702], water can only be determined if not more than 15% by volume are added to the working medium.

Sulphonic acids act like sulphuric acid and must be neutralized prior to a water determination. Lugowska et al. [7001] neutralize dodecylbenzosulphonic acid with pyridine and determine the water by a back-titration.

8 Inorganic Compounds

The determination of water is applied widely in inorganic compounds, mainly to crystalline salts and to liquid and gaseous acids. Problems can be presented by certain oxides and hydroxides which react with the KF solution.

Salts which crystallize without forming hydrates can contain adherent moisture and entrapped mother-liquor. Sodium chloride for example, can contain as much as 0.3% entrapped water which diffuses very slowly to the surface. At room temperatures this can take a few weeks, and several days in a drying cabinet at 105 °C. It is therefore not surprising that this water can not be determined by a KF titration if the substance is insoluble.

If the total water content is to be determined, the working conditions must be so chosen that the salt dissolves. If only the adherent moisture is to be determined, the salt must be prevented from dissolving, by diluting the working medium with chloroform for example.

The same generally applies to compounds which contain water of hydration. For a water titration, these salts should be completely dissolved if possible. But there are a number of hydrates, however, which release their water of crystallization sufficiently rapidly in suspension [4101]. Adherent water is sometimes more difficult to determine as the hydrate water is partially titrated as well. For fertilizers, dioxane is used to dissolve the adherent moisture [6405, 6909].

Strong acids have to be neutralized prior to a water determination. Soluble oxides, hydroxides and carbonates react with the KF reagent thereby producing water.

An indirect water determination can also be employed in order to avoid complications. Here, the water is either distilled off azeotropically and the distillate is titrated subsequently [6402], or evaporated in a tube furnace and carried into the titration cell by means of a stream of dry gas [6905, 7604].

8.1 Halides

Sodium chloride, ammonium chloride, potassium chloride, potassium iodide, cesium iodide and other alkaline halides crystallize without any water of hydration but usually contain included moisture. This can only be determined if the substance completely dissolves. A mixture of methanol and formamide can improve the solubility of the salts [Hyd. p. 37]. If only adherent moisture is to be determined, methanol/chloroform (1 + 3) is to be preferred as the working medium and the titration should be carried out as quickly as possible so as to prevent any noticeable dissolution of the sample.

The alkaline earth halides crystallize primarily as hydrates, for example magnesium chloride hexahydrate, calcium chloride dihydrate, strontium chloride hexahydrate or barium chloride dihydrate and the water as such can be titrated without any problems [4101]. The same applies for bromides and iodides.

The halides of other metals can also be analysed, for example aluminium chloride hexahydrate, ferrous chloride tetrahydrate, manganese chloride tetrahydrate, cobaltous chloride hexahydrate or stannic chloride pentahydrate [4101]. "False" results are given by cupric chloride dihydrate; only 1.5 moles of water are found because of a side reaction [4102]:

$$CuCl_2 \cdot 2H_2O + 1.5 I_2 + 2SO_2 + 2CH_3OH$$
$$\rightarrow CuI + 2HCl + 2HI + 2HSO_4CH_3. \qquad (8.1)$$

As this reaction takes place stoichiometrically, it can be allowed for in the calculations so that the water content can be determined. Van Acker et al. used a back-titration for the analysis of copper salts [7303]. Ferric chloride exhibits an anologous oxidation reaction which can be prevented by the addition of an excess of 8-hydroxyquinoline [5202]. Stannous chloride is a strong reducing agent and reduces the iodine in the KF solution [4102]. With titanium(III) chloride and vanadium(III) chloride, Giuffre [6402] distilled the water off azeotropically and titrated the distillate.

Soluble fluorides like potassium fluoride and potassium fluoride dihydrate can be titrated without any problems. Calcium fluoride is insoluble and only the adherent moisture can be determined. Potassium hydrogen fluoride dissolves only very slowly and Barbi and Pizzini recommend using acetic acid as the working medium [6305]. Titrating soluble fluorides, a titration vessel made of plastic should be used in order to prevent reactions with glass, which produce water.

Aqueous hydrochloric acid with the normal acid content of 37% can be titrated directly in a methanolic working medium. Because of a high water content, much KF reagent is required to reach the equivalence point. The amount of base thus introduced is sufficient to neutralize the acid. A buffered working medium in which the acid is immediately neutralized, would appear to be safer as it prevents the esterification of the acid.

Gaseous hydrochloric acid can also be adsorbed by methanol, though an immediate neutralization is then necessary. Milberger et al. [4903] adsorb the gas in a mixture of methanol and pyridine which contains an excess of KF reagent. The excess is then back-titrated. Muroi and Ono [7301] condense the water in a special apparatus at $-78\,°C$, separating most of the hydrochloric acid this way. The water thus frozen out is dissolved in methanol/pyridine and titrated. Gaseous hydrobromic acid is dissolved by Weintraub and Apelblat [8204] in pure pyridine and titrated.

Aqueous hydrofluoric acid can be analysed in the same way as hydrochloric acid. It should not be overlooked, however, that the substance is highly reactive. Glass vessels can only be used when the hydrofluoric acid has been neutralized with a suitable base. Anhydrous hydrofluoric acid (boiling point: $+19\,°C$) is neutralized prior to titration. According to ISO 3699, it is added to pyridine [7605] and titrated with a one-component reagent using special equipment.

8.2 Oxides, Hydroxides and Peroxides

Oxides and hydroxides react with acids producing a salt and an equivalent amount of water. The KF reagent behaves like a weak acid because it contains pyridinium salts of methyl sulphurous acid, methyl sulphuric acid and hydroiodic acid formed during the water titration. It reacts with basic substances in different ways. There are a few mineral oxides which are not dissolved by the KF solution. In these compounds the water content can be titrated without any disturbances. Other types of basic substances react rapidly and stoichiometrically with the KF solution. In this case the water produced can be eliminated by calculation. A few oxides react slowly with the KF solution, thus inhibiting a water determination.

Oxides and hydroxides which are easily soluble, react according to Eqs. (8.2) and (8.3). Other weakly basic compounds like carbonates and sulphites for example, react analogously to Eq. (8.4). They behave like a mixture of alkali oxide and carbon dioxide.

$$MgO + I_2 + SO_2 + CH_3OH \rightarrow MgI_2 + HSO_4CH_3. \tag{8.2}$$

$$NaOH + I_2 + SO_2 + CH_3OH \rightarrow NaI + HI + HSO_4CH_3. \tag{8.3}$$

$$K_2CO_3 + I_2 + SO_2 + CH_3OH \rightarrow 2\,KI + CO_2 + HSO_4CH_3. \tag{8.4}$$

MgO, CaO, Cu_2O, ZnO, HgO and Ag_2O react in accordance wit Eq. (8.2), but the rate of the reaction depends strongly on the structure of the oxides [4102]. Mineral oxides often react slowly and not completely. Lead oxide does not react completely because the crystals become covered with a layer of lead iodide which prevents any further reaction. Insoluble oxides like ferric oxide or aluminium oxide do not react at all, and adherent moisture can be titrated easily. The solubility can be influenced by the addition of acetic acid or of a (limited amount) of mineral acid. Manganese dioxide reacts according to Eq. (8.5) with the KF reagent. Lead oxides, PbO_2 and Pb_3O_4, react analogously although the reaction usually remains incomplete as the oxide particles become entrapped by lead iodide.

$$MnO_2 + I_2 + 2\,SO_2 + 2\,CH_3OH \rightarrow MnI_2 + 2\,HSO_4CH_3. \tag{8.5}$$

Hydrogen peroxide and related compounds react according to Eq. (8.6) with the sulphur dioxide present in the KF reagent, forming sulphuric acid or sulphates. Neither water nor iodine is either consumed or formed here, so that the water determination is not influenced at all. However, the acid formed must be neutralized.

$$H_2O_2 + SO_2 \rightarrow H_2SO_4. \tag{8.6}$$

Peroxides and persalts of metals react in a similar manner and the KF titration is not influenced. If the peroxides contain normal oxides or hydroxides, these by-products react according to Eqs. (8.2) or (8.3) producing water. Thus, oxides can be detected in peroxides by the KF titration. Na_2O_2, BaO_2, MgO_2, ZnO_2, CaO_2,

$(NH_4)_2S_2O_8$, $Na_2CO_3 \cdot H_2O_2$, $NaBO_2 \cdot 3H_2O \cdot H_2O_2$ and $Na_4P_2O_7 \cdot H_2O_2$ have been investigated [Aqu. p. 669].

Alkyl peroxides react according to Eq. (8.7). This reaction does not produce water and therefore does not affect the KF titration. In the absence of sulphur dioxide, iodide is oxidized according to Eq. (8.8).

$$ROOH + SO_2 \rightarrow HSO_4R. \tag{8.7}$$

$$ROOH + 2HI \rightarrow ROH + I_2 + H_2O. \tag{8.8}$$

Reaction (8.8) produces iodine and water, which are consumed by a later addition of sulphur dioxide without the KF titration being affected. The most important requirement for the analysis of peroxides is therefore a sufficiently high excess of sulphur dioxide in the KF system, as the KF reaction would otherwise be brought to a halt due to a lack of sulphur dioxide. Two-component reagents are particularly suitable for this as the amount of sulphur dioxide can be administered separately.

Dialkyl peroxides, R—OO—R, are inert and do not influence the KF titration. Diacyl peroxides, RCO—OO—OCR, oxidize the iodide according to Eq. (8.9). With benzoyl peroxide, this reaction appears to proceed so slowly that a water determination is possible. Short chained peroxides seem to be more reactive and it proves practical to cool these to 0 °C or to −20 °C [Aqu. p. 374–379].

$$RCO-OO-OCR + 2HI \rightarrow 2RCOOH + I_2. \tag{8.9}$$

8.3 Sulphur Compounds

For the determination of water in sulphuric acid, first a neutralization is necessary in order to prevent a shift of pH in the KF system and a possible esterification. The methanolic working medium is therefore supplemented with at least an equivalent amount of a suitable base (pyridine or imidazole) and pre-dehydrated. The sulphuric acid is then added and titrated as quickly as possible. Concentrated or fuming sulphuric acid should be neutralized in an alcohol-free medium, for example with cooled pyridine or mixtures of pyridine/tetrahydrofuran or pyridine/dioxane (1 + 1) [Aqu. p. 616).

The water content of sulphur dioxide had originally inspired Karl Fischer to the development of his reagent. Originally the sulphur dioxide had been visually titrated directly with the methanolic KF solution in a double-walled Dewar-vessel [Aqu. p. 598–601]. Di Caprio [4702] used pre-dehydrated methanol as the working medium, added the sulphur dioxide in liquid form and then titrated the water content. The water determination is simple to carry out; however, particular measures are required when taking samples because liquid sulphur dioxide will self-cool if not stored under pressure thus causing condensation of moisture.

Sulphates often crystallize as hydrates which can be dissolved in methanol easily and can therefore also be titrated well. Examples which can be mentioned are sodium sulphate decahydrate, magnesium sulphate heptahydrate, manganese sulphate tetrahydrate, nickel sulphate hexahydrate, zinc sulphate heptahydrate [4101], aluminium sulphate octodecahydrate, potassium aluminium sulphate dodecahydrate and ammonium ferrous sulphate hexahydrate. On the other hand, a water determination in zinc sulphate monohydrate or magnesium sulphate monohydrate is not possible as these substances are insoluble in methanol and the water is too tightly bound. Cupric sulphate pentahydrate acts in an oxidizing manner similar to cupric chloride (see Chap. 8.1) and it appears to contain only 4.5 moles of water. Sometimes it can cause indication errors as the electrodes are depolarized and the end point is indicated before starting the titration.

Calcium sulphate dihydrate releases its hydrate water so slowly that a KF titration at normal temperatures is not possible. Calcium sulphate hemihydrate contains the water less tightly bound and can be titrated within a reasonable period of time. The dihydrate is completely dehydrated at temperatures above 200 °C in a tubular furnace; Rechenberg [7604] used this substance to standardize the KF solutions used for the tubular furnace technique.

Hydrogen sulphates act as mixtures of neutral sulphate and sulphuric acid. When using larger amounts, a prior neutralization is necessary.

Sodium sulphite Na_2SO_3, sodium disulphite $Na_2S_2O_5$, and also sodium thiosulphate $Na_2S_2O_3$, react quantitatively with iodine and this led to the proposal by Johansson [5601] for example, of first oxidizing the thiosulphate with iodine in a methanol/pyridine solution and then adding the sulphite/pyridine component before determining the water content with the iodine solution.

Sodium dithionite $Na_2S_2O_4$, does not react with iodine in methanolic solutions, and water can thus be determined.

Hydrogen sulphide and soluble sulphides are oxidized to sulphur whereby the reaction does not appear to remain exactly stoichiometric.

8.4 Selenium and Tellurium Compounds

The moisture content or water of crystallization in alkali metal salts of selenium and tellurium can be determined accurately. Sodium selenate behaves like sodium sulphate and the water content can be titrated directly. According to Chiang and Cooper [6304], sodium selenite, potassium tellurite and sodium tellurate behave like a mixture of the respective oxides. The alkali metal oxide in these compounds reacts quantitatively with the KF solution. The water produced can be calculated. Selenites are reduced by the sulphur dioxide. The complete reaction is described by Eq. (8.10).

Tellurites and tellurates are reduced in the same manner as selenites.

$$Na_2SeO_3 + I_2 + 3SO_2 + 3CH_3OH$$
$$\rightarrow 2NaI + Se + 3HSO_4CH_3. \tag{8.10}$$

8.5 Nitrogen Compounds

Just like every strong acid, nitric acid has to be neutralized prior to a water determination. 100% nitric acid can also be analysed. Moberg et al. [5602] found that a contents of free NO_2 up to 1.5% did not cause any detrimental effects. For red fuming nitric acid containing up to 15% NO_2, a special working technique was employed by Burns and Muraca [6306] and corrective values were calculated for the NO_2 content.

Nitrates are easily analysed. Ammonium nitrate [Eb. p. 55] is referenced in the literature as are cobaltous nitrate hexahydrate, mercurous nitrate monohydrate and chromium(III)-nitrate-9-hydrate [Aqu. p. 88–89]. According to Mitchell et al. [4102], sodium nitrite reacts with the KF reagent according to Eq. (8.11).

$$NaNO_2 + 0.5\,I_2 + SO_2 + CH_3OH \rightarrow NaI + NO + HSO_4CH_3. \tag{8.11}$$

Ammonia can not be titrated directly as it reacts too alkaline. A neutralization using acetic acid or benzoic acid is adequate in order to titrate aqueous ammonia solutions. Liquid ammonia contains only a very small amount of water, about 100 ppm, and considerable amounts have to be applied. As much of the ammonia as possible must therefore be removed before determining the water content. Hodgson and Glover [5203] added ethylene glycol to liquid ammonia at $-78\,°C$ and then allowed the ammonia to evaporate upon warming. The water is retained by the ethylene glycol and can be subsequently titrated. A similar procedure is recommended by ISO/DIS 7105 for liquefied ammonia [8305].

Hydroxylamine can be oxidized by the iodine and therefore affects the KF titration [Aqu. p. 681]. Adding an excess of sulphur dioxide, amidosulphuric acid is formed according to Eq. (8.12). A water determination using a two-component reagent is therefore better suited as it contains an excess of sulphur dioxide [Hyd. p. 36].

$$H_2NOH + SO_2 \rightarrow H_2NSO_3H. \tag{8.12}$$

Hydrazine derivatives show very differing behaviours. Whereas hydrazine sulphate can be analysed without any difficulty, hydrazine hydrate and hydrazine hydrochloride consume a stoichiometric amount of iodine. These substances can also be analysed though in a sufficiently acidic solution, for example by using a two-component reagent in the presence of a stoichiometric amount of sulphuric acid [Hyd. p. 36].

8.6 Phosphorus Compounds

The determination of water in phosphoric acid presents no problems provided it has been neutralized beforehand. Primary phosphates can be titrated directly [4101]. Soluble secondary phosphates react slightly alkaline (pH 9). They become neutralized during the course of the titration by the acids formed and therefore do not present any problems either. Tertiary phosphates react strongly alkaline and must be neutralized with acid prior to determining the water. Tri-sodium phosphate dodecahydrate normally contains approximately 0.2 moles of free NaOH, which is also neutralized during the KF titration [Aqu. p. 680]. Magnesium hydrogen phosphate trihydrate, magnesium ammonium hydrogen phosphate hexahydrate and primary and secondary calcium phosphates can also be analysed. Secondary calcium phosphate, however, which is insoluble, releases water only very slowly. Fertilizers containing phosphates are covered in Section 10.11.

The esters of phosphoric acid, like tributyl phosphate, can also be analysed, whereas phosphines are oxidized by the iodine [Aqu. p. 648]. Hypophosphite is not oxidized by the KF solution, so that the water content of these compounds can be determined by titration.

8.7 Arsenic and Antimony Compounds

According to Mitchell at el. [4102], sodium arsenite acts like a mixture of sodium oxide and arsenic(III)oxide, i.e. both oxides react analogously to Eq. (8.2). The total reaction is expressed by Eq. (8.13). Arsenates react analogously whereby the 5-valent arsenic is also reduced. The course of the reaction is given in Eq. (8.14).

$$NaAsO_2 + 2I_2 + 2SO_2 + 2CH_3OH$$
$$\rightarrow NaI + AsI_3 + 2HSO_4CH_3. \tag{8.13}$$

$$Na_2HAsO_4 + 3I_2 + 4SO_2 + 4CH_3OH$$
$$\rightarrow 2NaI + AsI_3 + HI + 4HSO_4CH_3. \tag{8.14}$$

Antimony(III)oxide dissolves slowly in the KF reagent under titration conditions, so that only adherent moisture can be titrated if 2-methoxyethanol is used as the working medium. A rapid titration is to be preferred. Antimony(III)chloride can also be investigated, whereby the addition of a small amount of sulphuric acid to the titration medium helps to prevent a possible hydrolysis.

8.8 Carbonates, Bicarbonates

Both these classes of compounds react with KF solutions, carbonates according to Eq. (8.4) and bicarbonates according to Eq. (8.15). The rate of reaction depends to a large extent on the solubility. Alkali carbonates react rapidly, alkaline earth carbonates relatively slowly and bismuth carbonate shows no reaction at all.

$$KHCO_3 + I_2 + SO_2 + CH_3OH \rightarrow KI + HI + CO_2 + HSO_4CH_3. \tag{8.15}$$

Water determinations in these compounds are only possible by extracting with methanol or other solvents. For sodium bicarbonate, which is soluble in methanol to a certain degree, the amount dissolved should be titrated acidimetrically and taken into account in the calculation. Drying in a tubular furnace and subsequent KF titration of the evaporated water is also a suitable method.

8.9 Silicon Compounds

Silica gel is an important desiccant whose water content can be determined [4101] when it is pulverized. The surface moisture can also be determined on precipitated or pyrogenically manufactured silica [6706]. Alkali silicates react as a mixture of alkali oxide and silicon dioxide, i.e. the alkali oxide forms water with the KF reagent.

Organosilicon compounds also have been investigated. Apparently siloxanes can be titrated without any difficulty [Aqu. p. 321]. According to Smith and Kellum [6603], silanols R_3SiOH, react with methanol to form ethers according to Eq. (8.16).

$$R_3SiOH + CH_3OH \rightarrow R_3SiOCH_3 + H_2O. \tag{8.16}$$

This reaction occurs so rapidly that the KF titration is affected. The reactivity of the individual silanols is very different. Kellum and Smith [6705] classified the reactivity of the triorganosilanols as follows:

$$(CH_3)_3SiOH \gg (C_6H_5)(CH_3)_2SiOH > (C_6H_5)_2CH_3SiOH > (C_6H_5)_3SiOH.$$

Smith and Kellum [6603] recommend using long-chain alcohols as the working medium for the determination of water in these compounds; for example, mixtures of 2-ethylhexanol or dodecanol with pyridine. A further recommendation for monomeric or short-chain silanols is to carry out the titration at low temperatures [6705], 0 °C for example.

Kellum and Smith [6706] also applied the KF titration to the determination of free silanol groups and reactive siloxane compounds in which they catalyze the reaction to an ether as in Eq. (8.16) and the cleavage of the siloxane bonds by boron trifluoride.

Mika and Čadersky [7203] investigated the coulometric determination of water in silanols. They used 2-methoxyethanol as the solvent thus inhibiting the formation of the ether as given in Eq. (8.16).

Using the newly developed Hydranal-Composite 5K for the determination of water in ketones the silanols can be analysed for water easily.

8.10 Boron Compounds

Boric acid reacts with the alcohol in a KF reagent causing esterification according to Eq. (8.17). Boron trioxide B_2O_3, metaboric acid HBO_2, and sodium tetraborate also act in the same way [4102]. The rate of esterification decreases with increasing chain length of the alcohol used.

$$H_3BO_3 + 3CH_3OH \rightarrow B(OCH_3)_3 + 3H_2O. \tag{8.17}$$

9 Foodstuffs

The term foodstuffs comprises substances of very differing structures. It includes well-defined chemical substances like glucose or saccharose as well as simple mixtures like alcoholic beverages, but also starch, flour, dried fruits and meat which have a very complex cellular structure.

The water is bound very different in these products. Schneider depicted the forms of inclusion in vegetable solids in Fig. 9–1. Some water is sorptively bound to the surface of the grains. Further amounts of water are bound in the capillaries, macrocapillarily between the individual grains, and also microcapillarily in the grain capillaries themselves. Furthermore, water is also entrapped in the cells. The vapour pressure of the water depends on the type of water inclusion. Sorptively bound water is more easy released than is capillarily bound. Schneider [8114] clarified the relations by means of

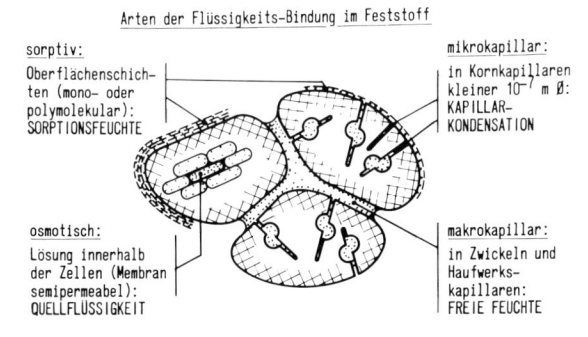

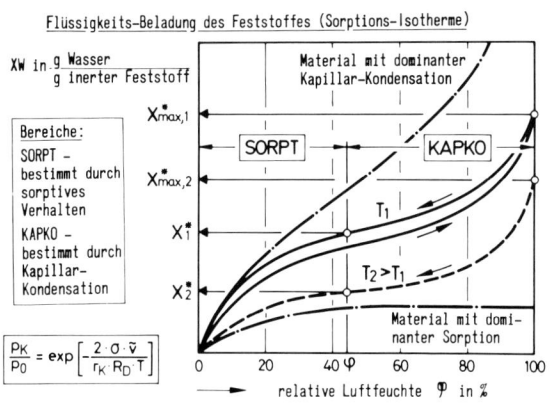

Fig. 9-1 Forms of inclusion of water in vegetable solids according to Schneider [8114]. (With kind permission of the author)

adsorption isotherms (see Fig. 9–1). If the relative humidity in the external air is decreased, as in a dessicator for example, the substance will release water. If the temperature is raised, another absorption isotherm then applies, and the substance releases water, too. It always requires a definite length of time before a new state of equilibrium is reached. In certain cases, this period of time can be very long, so that it can not be made use of for a practical determination of water. It should also be taken into account that a release of water sometimes results in an irreversible change of the dried substance.

It does not surprise that the most usual methods for water determination like oven drying, vacuum drying or azeotropic distillation led to conflicting results. This had been shown by investigations on starch products by Eberius and Kempf [5703] or on dried apricots by Zürcher and Hadorn [7801]. The results are influenced to a large extent by the test conditions, i.e. by drying temperature, length of drying time, particle size of the sample and the mode of heat transfer. Results can also be influenced by loss of chemically bound water or by slowly occurring condensation reactions in which water is formed. The sample can also contain other volatiles which can evaporate at the same time thus implying a higher water content. For the practical determination of water, therefore, it has become common practice to standardize the methods of drying and prescribe very detailed procedures.

The Karl Fischer titration appears to fare somewhat better with these difficulties. Adherent moisture and capillary-bound water are easily accessible and a water titration takes only a few minutes. Side reactions of a chemical nature can usually not occur as the known classes of reactive substances like aldehydes and mercaptans are hardly present in foodstuffs. As the KF titration is carried out at room temperature or at slightly elevated temperatures, and as it takes only a few minutes, thermal decomposition of the substance is not to be allowed for. Cellular moisture can cause problems, as the extraction proceeds very slowly. It can take several days at room temperature and titrations at higher temperatures are often recommended. The water can be made easily accessible if the substance is triturated finely so that the cells are ruptured [7903].

The water content found by Karl Fischer titration is often compared with the loss on drying in order to check the reliability of the Karl Fischer titration. A few investigations should be cited: Mitchell studied cellulose [4002]; Johnson sugar, starch and dried vegetables [4501]; Eberius and Kempf, starch products [5703]; Sandell, carbohydrates [6001]; Epps, molasses [6602]; Martin, chocolates [7707, 7708]; Zürcher and Hadorn, flour, margarine, corn starch and dried apricots [7801]. The results, achieved by the collaborative study on chocolate, are listed in Table 9–1. They can be considered as typical: the water contents found by Karl Fischer titration are different from those obtained by oven drying, and the reproducibility of the KF values is better.

Many other comparison investigations confirm this result. Normally the water content found by Karl Fischer titration is somewhat higher than the loss on drying. This difference seems to be reasonable when we consider that the water is present in different forms and it is analysed by different methods. We should therefore keep the terms "loss on drying" and "water content by Karl Fischer titration" distinctly apart.

In spite of its incontestable advantages, the KF titration is applied to the investigation of foodstuffs only to a certain degree. This may be explained by the fact that with certain foodstuffs the titration is sluggish and apparently no end point is reached in a

9.1 Handling Samples

Table 9-1. Statistical analysis of data from a collaborative study on chocolate by Martin et al. [7708]

	Sample				
	1	2	3	4	5
	Karl Fischer				
Mean	1.20	1.21	0.48	0.87	1.20
Reproducibility	0.060	0.084	0.070	0.039	0.052
Coeff. of var., %	5.0	7.0	14.6	4.5	4.4
Repeatability	0.011	0.019	0.019	0.013	0.012
Coeff. of var., %	0.9	1.6	4.0	1.5	1.0
	Oven				
Mean	1.33	1.20	0.58	0.86	1.29
Reproducibility	0.277	0.256	0.120	0.187	0.241
Coeff. of var., %	20.8	21.3	20.6	21.7	18.6
Repeatability	0.067	0.065	0.035	0.033	0.045
Coeff. of var., %	5.0	5.4	6.0	3.8	3.5

reasonable time. This does not create the impression of reliability. Titration at higher temperatures, as recommended by Zürcher and Hadorn [7801, 7903, 8101], and the use of pyridine-free reagents [8302] have decisively altered this point of view. Under these conditions, titrations mostly can be carried out in less than 10 minutes and the results are more accurate than those obtained by conventional methods. The application of the Karl Fischer titration to foodstuffs will arise in the future. In 1982, the "Schweizerische Lebensmittelbuch" included the KF titration as a standard method [8207].

9.1 Handling Samples

Foodstuffs are not uniform substances and many are either insoluble or only partially soluble and samples must therefore be handled on an individual basis. Specific details are given with each substance covered, but several general aspects should be mentioned beforehand.

Firstly, the possible inhomogenity of the substance to be investigated must be mentioned. When bread is analysed, the outside crust certainly has a different water content than that of the interior of the loaf, and when investigating plant or animal tissue, the water content is never distributed homogeneously. A sample must be so chosen that it represents the actual average. It must then be homogenized, by grinding for example [7804], so that the portion taken for the analysis is truely representative of the whole.

The sample must be further processed so that it will release the water contained within an acceptable period of time. It is generally common to reduce the particle size of the substance such that it will pass through a 40 mesh sieve [4501, 5802, 5803]. If the sample contains cellular tissue, the cells should be ruptured if possible in order to ease the release of water. High performance mixers like the Hydractor for example [Aqu.

p. 499], a mixer with 15,000–18,000 rpm, are to be recommended for this. Also, the so-called "high frequency shredding" [7804, 7903, 8207] from Zürcher and Hadorn is a method of trituration. A Polytron homogenizer model PT 20 TS is used, whose stator and rotor are saw-toothed matched and can achieve high sheer strengths at 20,000 rpm. The sample is adequately triturated within 2–5 min. A further method is wet milling in methanol as proposed by Jones [8105, 8107] for the investigation of wheat. He used a 65 ml milling vessel made of stainless steel with metal milling pellets and a milling time of 30 min.

As long as the substances are soluble, as is the case with certain sugars, hardly any problems are encountered with the water determination. Only a suitable solvent has to be found to dissolve the sample rapidly. The working conditions should be optimized so as to prevent side reactions taking place.

Most samples are insoluble, however, and the water has to be extracted. Methanol is usually used as the extraction agent because it extracts water more quickly than other alcohols. Glycol or even formamide are used in exceptional cases, or mixtures of solvents containing glycol or formamide. The extraction can take 30–90 min at room temperature. Sometimes longer extraction times of up to 20 h may be required. In exceptional cases 2 days can be necessary, as reported for dried vegetables [4802]. Extractions are often carried out at elevated temperatures whereby the extraction times can be considerably reduced, to 30 min for example for starch and dried vegetables [4501, 7205]. The extraction can sometimes remain incomplete if the sample is too large. Methanol apparently loses its capacity for further extraction if it contains more than 0.5% water [7801].

Extraction by refluxing with methanol is also a recommended method and an extraction time of 15 min appears adequate for starch products [4802]. Sometimes higher temperatures or other solvents are to be recommended, like formamide for example, which has a boiling point of 193 °C. This method requires confidence in handling as it is not without a number of risks. Due to the somewhat complicated technique, moisture from outside can easily infiltrate and offset the results. Because of the high temperatures involved, chemical decomposition or even dehydration can take place as seen in oven-drying methods with fructose and other sugars for example. Extraction using dioxane is also employed in conjunction with the azeotropic distillation of water as in the study of coffee beans for example [7709].

9.2 Titration Techniques

The titration of extracted water or of substances soluble in methanol can be carried out using standard procedures without any limitations. Side effects of a chemical nature are not usually to be expected as only in exceptional cases reactive components (ascorbic acid) are present in appreciable concentrations in foodstuffs.

A back-titration is sometimes recommended when an end point is too dragging with a direct titration, and a back-titration can give a more definite end point (refer to Section 3.4). This argument is only apparently applicable. A fading end point during a direct titration indicates that the extraction of the water is not yet complete. By reversing the mode of titration, the problem of an incomplete extraction of water is rather ignored than solved. If the direct titration is carried out perfectly, a back-titration will not bring any further advantages [7801]. It only increases the amount of work involved and should only be employed as a well-founded exception.

A KF titration at a higher temperature is an efficient combination of the extraction at a higher temperature and the direct water titration (Section 3.5). By this method, one can dispense with a separate extraction apparatus and hence reduce the errors which can be caused by foreign moisture. Also, the working technique is simplified. In practice, a titration cell with a water jacket is used and this is connected to a thermostat. With several KF titrators, a temperature-controlled hot plate can be placed under the titration cell and the temperature can be maintained easily. If methanol is used as the working medium, one can heat up to 50 °C without having to employ any additional apparatus like a condenser. If a heating temperature of 70 °C is chosen, as done by Zürcher and Hadorn [8101], the methanol will boil and the cell must be equipped with a refluxing condenser which then has to be separated from the outside surroundings by a drying tower. The application of this working technique to approximately 100 single products is fully described by Zürcher and Hadorn [8101].

We found the use of pyridine-free Hydranal reagents to be particularly advantageous [8302] as the titration times were thereby shortened to 2–10 min especially when titrating at elevated temperatures.

9.3 Carbohydrates

In this section not only pure carbohydrates shall be discussed but also several products which consists mainly of carbohydrates. As in the case with all foodstuffs, it is difficult to fix the limits of other foodstuffs.

9.3.1 Sugars and Sugar Products

The determination of water in different sugars is possible if the working conditions are compatible with the properties of the individual substance. Basically the sugar has to completely dissolve if the total water content is to be determined. Several substances contain water of hydration or a non-stoichiometric amount of water trapped in their crystal lattice (presumably in liquid pockets), which does not diffuse out during titration. In the solid state, known hydrates like lactose monohydrate for example, release their water of crystallization also very slowly. The working medium must therefore be chosen so that the substance will dissolve completely, and rapidly if possible. McComb [5701] recommends methanol/formamide mixtures for the analysis of maltose and lactose. Muroi, Tsutsumi and Koizumi [6906] used the same mixture for candy and caramel products. It was also recommended to carry out the titrations at elevated temperatures even though decompositions of the sugars can take place. Johnson [4501] as well as Zürcher and Hadorn [8106] found extensive decomposition of fructose at 65 °C. We found galactose to decompose slightly.

D(+)-glucose dissolves in methanol and in Hydranal-Solvent at 20 °C and is easily titrated and a theoretical water content of 9,09% has been found [8302]. It also can be titrated at 50 °C. At 70 °C, a vanishing end point indicates the beginning of decomposition. Zürcher and Hadorn also found a slight decomposition at 65 °C [8106].

D(+)-galactose is only sufficiently soluble (0.5 g/20 ml) in formamide/methanol or in formamide/Hydranal-Solvent. 0.20% water had been found at 20 °C [8302]. This sugar begins to decompose at 50 °C and an end point can not be achieved at this temperature. In methanol, galactose is only very slightly soluble (approx. 0.1 g/20 ml). The water included in the crystals is not extracted during titration. Therefore, we found only 0.01% of water for a sample weight of 5 g, corresponding to the amount of galactose which had dissolved. A decomposition at 50 °C is noticable in methanol as well, even though the substance does not completely dissolve.

D(−)-fructose exhibits good solubility (1 g/20 ml) at 20 °C in formamide/methanol (1 + 1) and the water content (0.06%) can be easily titrated. We achieved stable end points at temperatures up to 70 °C as well, without any indication of decomposition [8302]. The solubility is adequate in methanol or Hydranal-Solvent at 50 °C so that these solvents can also be used.

D(+)-maltose monohydrate dissolves sufficiently in methanolic solutions at 20 °C, but formamide mixtures are also suitable. We obtained slightly vanishing end points at 70 °C, the first indication of decomposition [8302]. We found 5.85% water, the calculated value for the monohydrate is 5.00%. The substance contains more than one mole of water as McComb [5701] had previously observed. She found approximately 5.60% water corresponding to 1.12 mole. Only 1.7% of water was found by oven drying (3 h at 103 °C).

D(+)-lactose monohydrate dissolves very slowly at 20 °C and incompletely in methanol and therefore gives a sluggish titration and values which are too low. The substance dissolves well in formamide/methanol (1 + 1). Titrating at higher temperatures, methanol is also satisfactory and the water content can be quantitatively ascertained. According to Thomasow, Mrowetz and Delfs [7202], the water is released very slowly and incompletely by oven drying at 102 °C, which indicates that loss on drying and water content cannot agree. Similar differences between KF titration and loss on drying must be expected for foodstuffs containing lactose, such as milk and milk products.

Saccharose has been analysed for both total water content and adherent moisture. The adherent moisture can be determined if the sugar can be prevented from dissolving by using a saturated methanolic solution as the working medium for example [7505]. We prefer to use a mixture of 20 ml chloroform and 5 ml Hydranal-Solvent [8302]. 5 g Saccharose are suspended in this solvent mixture and titrated rapidly. The titration takes less than one minute. Working without the addition of chloroform, the saccharose will slowly dissolve and a higher water content is found which is proportional to the amount of dissolved sugar. Saccharose contains a certain amount of water included in its crystals. In the course of weeks and months, this moisture diffuses out of the crystal and causes the sugar to "become moist". These amounts can only be ascertained by a water determination when the substance is fully dissolved. Schneider, Emmerich and Ticmanis [7102] found that it can take approximately one hour to dissolve the sample in formamide. Because of the long titration time, a "drift" compensation is necessary in order to eliminate the blank consumption of the cell during this time. Saccharose dissolves rapidly and in sufficient amounts in formamide/Hydranal-Solvent (1 + 1) at 50 °C to 70 °C and KF titration can be carried out within 3 min [8302].

Molasses can be titrated in a similar way. Epps [6602] found good agreement in his collaborative study between loss on drying and water content according to KF, though the KF values were generally higher, especially for "heavy cane" molasses which only contained approximately 19% of water. According to Kviesitis [7704], problemes were encountered with the dissolution in methanol. As well as methanol, formamide and mixtures of the two were tried [7805]. No matter which solvent was used, it always took up to 10 min to dissolve the molasses. A detailed working procedure is presented in "Sugar Analysis" published by ICUMSA in 1979. According to this method, the molasses are injected into pre-dehydrated methanol using a syringe and then titrated. A formamide/methanol mixture and a working temperature of 50 °C would also appear to be very practical here.

Candy and soft caramel products usually exhibit a very unsatisfactory solubility and this led Sandell to propose an extraction with methanol at 65 °C [6001]. Muroi et al. dissolved and titrated the caramel mass at 55–60 °C in formamide/methanol [6906]. We prefer a titration at 50 °C in formamide/methanol. Under these conditions, the substance dissolves rapidly and completely [8302].

Honey dissolves in methanol whereby stirring for a few minutes will suffice. Also, there are no problems with the subsequent titration [8101].

9.3.2 Starch, Dextrins, Pectins, Flour

These carbohydrates are usually insoluble because of their high molecular weight and extraction techniques have to be applied.

Johnson [4501] extracted potatoe and grain starch for 30 min with boiling methanol and titrated the extracts. According to Sandell [6001], heating to 60 °C for a short time is sufficient to complete the extraction of corn starch.

Eberius and Kempf [5703] found the water content of starch to be somewhat too low using a direct titration method, because the water does not diffuse out of the grains within the short titration time. They therefore recommended extracting potatoe and corn starch for 30 min at room temperature. The extraction time depends on the grain size and also on the hardness of the grain so that it can sometimes prove practical to extract at elevated temperatures as well [7205].

Zürcher and Hadorn [7801] found a direct titration of corn starch at room temperature to be possible if the cut-off of the KF instrument can be sufficiently delayed so that the titration takes 15 min. Agreeing results are given by a direct titration at elevated temperatures as well as by back-titration. They obtained somewhat lower values from a reflux extraction and even lower values by oven drying. Good agreement was reached in a collaborative study of warm and cold titrations [8008].

We found values somewhat too low (14.5%) for potatoe starch during a 15-min titration. The titration at 50 °C with Hydranal-Solvent/Hydranal-Titrant seems to be more advantageous as it takes only 4–5 min and the results are accurate (14.61–14.67% titrating the same potatoe starch). The same results were obtained by extraction with methanol for one hour at room temperature.

According to ISO DIS 5381, starch hydrolysis products are suspended in a mixture of methanol and formamide (7 + 3) and titrated directly [8206].

The grain size of flour is larger than that of starch and longer extraction times are required. Sandell [6001] recommended heating to 60 °C for a short period of time. A direct titration at room temperature starts off rapidly but a final value is reached only after 60 min [7801]. At elevated temperatures, the overall course of the reaction is quicker and higher water contents are found. The accuracy of these methods of determination was evaluated in a collaborative study [8008]. We found a titration time of 20 min to be necessary for wheat flour using the Hydranal reagent. At 50 °C the titration took a mere 5 min. In both cases we found the same water content (13.95%). For extraction with methanol one hour was adequate [8302].

A decisive factor for the titration time is the grain size. Wheat groats can not be titrated at room temperature, as even a titration time of one hour does not result in an end point. At 50 °C the titration takes 15–20 min and an extraction with methanol requires a minimum of three hours. If the wheat groats are finely ground in a mill, it resembles flour, as might be expected.

Pectins are similarly extracted with warm methanol, 15 min at 60 °C [4501], or by refluxing for 15 min [6001].

Because of its glassy structure, gelatine releases water very slowly and it is usually extracted with boiling methanol [6001] for approximately 30 min. The extraction can be accelerated by using a slight excess of KF reagent and titrating back. Gelatine is soluble (1 h) in a formamide/methanol mixture (4 + 1) and can equally be titrated this way.

9.4 Fats

The water content of fats can be determined without any difficulties if a suitable solvent is chosen which enables a complete and if possible a rapid dissolution of the substance.

Edible oils usually contain less than 0.1% water and large sample weights (5–10 g) are thus necessary. Chloroform is therefore to be recommended as the solution agent [4801]. Using conventional reagents, a titration takes approximately 10 min [8101], with Hydranal-Solvent/chloroform (2 + 1) only about one minute [8302].

A mixture of solvents containing chloroform is also to be preferred for butter to dissolve the fats. Administration of the sample requires particular care as the water in butter is not homogeneously distributed. Zürcher and Hadorn [8101] recommend the preparation of a stock solution in methanol/chloroform (1 + 1). We also found differing values within the stock solution, as gelatineous precipitations formed on the walls of the flask [8302]. We therefore prefer to titrate three or four samples directly. In order to weigh out the samples, we use a tared balance spatula made of PTFE (hydrophobic) which is then immersed with the sample into the working medium and which remains in it until the titration is complete.

Run butter, which contains less than 1% water, can be titrated directly using a sample weight of 5–10 g as prescribed in an international standard [6606].

Margarine can be analysed in exactly the same way as butter. The water appears to be more homogeneously distributed here, so that a plastic syringe is also suitable for weighing out the samples [8101].

Hard fats are best titrated at elevated temperatures as they dissolve fast in methanol/chloroform (1 + 1) at 50 °C. They can be directly weighed out using a PTFE spatula [8302] or by means of a syringe if the fat had previously been melted in a drying oven at 40 °C [8101].

9.5 High-Protein Foods

As expected, the determination of water in milk can be carried out without any disturbing effects [8101]. Special accuracy is required when weighing out the sample as the water is the main constituent. Condensed milk must first be homogenized, if possible by means of "high-frequency shredding" [8101].

Milk powder is extracted for one hour with methanol according to della Monica and Holden [6802]. An aliquote portion of the extract is then titrated. Thomasow, Mrowetz and Delfs [7202] titrated low-fat milk powder, standard milk powder and whey powder directly with KF solutions and compared the results with those of oven drying. Normally the loss on drying values are lower, as the dried products still contain water. For whey powder, the values obtained by oven drying are too high, because the substances decompose in the drying cabinet. Zürcher and Hadorn [8101] recommend the "high-frequency shredding" in order to prevent the formation of lumps. The titration takes only 10 min. With Hydranal-Solvent/Hydranal-Titrant, the titration is finished within 5 min. For standard milk powder, an addition of chloroform is an aid to dissolve existing fats [8302].

Joghurts and curds must be homogenized very thoroughly prior to titration and preferably using a good stirrer or a mixer.

For cheeses, large samples must be weighed out because of the inhomogeneous distribution of the water. Strange [6901] therefore extracted 5 g of cheese with n-butanol and titrated an aliquote portion thereof. In later investigations [7008] the sample was extracted in a mixer using anhydrous 2-propanol as a solvent. The results of a collaborative study of 4 sorts of cheeses in 5 laboratories proved satisfactory [7206]. Christen and Richardson [7909] homogenized the cheese directly in the titration cell in methanol/chloroform using a high performance stirrer (Polytron PT 10) but found an unsatisfactory accuracy. Zürcher and Hadorn obtained very highly reproducible results using the "high-frequency shredding" (Polytron PT 20 TS) [8101]. The titration takes approximately 20 min. Using Hydranal-Solvent/formamide as the working medium, the titration only takes 5–10 min at 50 °C [8302].

9.6 Various Foodstuffs

9.6.1 Fruit Juices, Alcoholic Beverages

Fruit juice, grape juice, vegetable juices and fruit juice concentrates do not present any problems to the water determination [8101]. They require accurate administration of the sample however, as they consist mainly of water. The same applies for alcoholic beverages. For high-proof alcohol (more than 99%), KF titration is the most reliable method of water determination.

9.6.2 Mayonnaise

A water determination takes 15–30 min using methanol as the working medium and a conventional KF reagent. It can be carried out within two minutes using Hydranal-Solvent/chloroform (2 + 1).

9.6.3 Baby Food

Baby food in powdered form is treated like milk powder. For higher sugar contents, formamide can improve the distribution.

9.6.4 Coffee, Instant Coffee

According to DIN 10 766 coffee can be ground and extracted for 20 min with dioxane. The water is azeotropically distilled off and the distillate is titrated with the KF solution [7709]. The determination of water in spray-dried or freeze-dried coffee can cause problems as the titration is sluggish [8101]. The problem can be overcome by choosing Hydranal-Solvent as the working medium and adding formamide as dispersion aid [8302].

9.6.5 Egg Powder

Zürcher and Hadorn [8101] found the water content to be too low for titrating directly, as lumps are formed. They therefore recommend to use "high-frequency shredding".

9.6.6 Pudding Powder

Good results are achieved by titrating at elevated temperatures. At room temperature the titration is too slow [8101].

9.6.7 Cocoa Powder, Cocoa Beans

Cocoa powder is best titrated in Hydranal-Solvent/formamide (1 + 1) at 50 °C. Cocoa beans should be finely ground beforehand and than treated in the same manner [8302].

9.6.8 Chocolate

Chocolate is relatively simple to analyse, provided a solvent is chosen which enables a rapid distribution of the sample. Kreiser and Martin [7707] used a mixture of methanol/formamide/chloroform (3 + 4 + 3). They found the KF results to be more accurate than those of conventional methods of determination. The results were confirmed by a collaborative study (see Table 9–1). The high-frequency shredding was also tested, but the titration times of 25 min were relatively long. We found a mixture of Hydranal-Solvent/chloroform (2 + 1) to be well suited [8302]. Using this working medium, milk chocolate could be titrated at room temperature. Chocolates which dissolve less readily can be analysed at 50 °C.

9.6.9 Almonds, Hazelnuts

A good reduction of particle size is a pre-requisite for a swift water determination here as well. Zürcher and Hadorn [8101] achieved titration times of 20–30 min at room temperature after triturating the sample by high-frequence shredding. We ground the sample using a laboratory grinder and titrated at 50 °C. Using Hydranal-Solvent as a working medium, the titration took about 10 min.

9.6.10 Marzipan

The same titration as for almonds can be performed. An addition of formamide proves useful as this improves the distribution [8302].

9.6.11 Pastries

A number of pastries were tested by Zürcher and Hadorn [8101]. Rusks are milled in a ball mill and titrated at elevated temperatures. High sugar content biscuits ("Petit-Beurre"), gave water contents which are too low after beeing ground in the ball mill and titrated at room temperature. No end point was reached with titrations at elevated temperatures because of chemical decomposition of the sugar contained. Milling at room temperature in methanol still only releases the water very slowly. Using Hydranal-Solvent/formamide as the working medium, the titration took about 15 min at room temperature [8302]. The analysis of bread takes relatively long and involves a lot of work because of the large sample required. A quarter of a loaf is pre-dried, milled and mixed well. A sample thereof is titrated. Even at elevated temperatures the titration is very sluggish.

9.6.12 Pasta

Noodles are very finely milled in a ball mill. The water can then be determined by a direct titration at elevated temperatures [5803]. In Hydranal-Solvent/formamide (2 + 1) the titration is completed within 10 min at 50 °C [8302].

9.6.13 Grain

Wheat, rye, rice, corn and other types of grain have been analysed. Seibel and Bolling [5802, 5803] tested many methods of grinding and extraction. They found the best method to be the dry milling and subsequent KF titration at 50 °C. Jones and Brickenkamp [8107] put the sample, milling balls and 50 ml of methanol into the milling vessel. The milling time is about 15 min, during which time the mixture is heated by milling just below the boiling point of methanol. After the residue has settled, an aliquote portion of the extract is titrated volumetrically or coulometrically. Zürcher and Hadorn [8101] milled the samples in a ball mill without methanol and then titrated at an elevated temperature.

9.6.14 Dried Vegetables, Dried Fruit

The water determination in dried vegetable poses problems. The water in the cell is only released slowly which also causes difficulties with oven drying methods. Johnson [4501] analysed many sorts of vegetables by extracting the samples with methanol at 60 °C and subsequently titrating the extract. An extraction time of 30 min usually gave an acceptable result though longer extraction times (6 h) did lead to a slight increase of the water amount found. Thung [6403] extracted vegetables with formamide at 70 °C. An extraction time of 30 min gave reliable results. This time dependance is more strongly reflected by the vacuum drying method. A continuous loss of weight has been found during a few weeks. Rader reported on a collaborative study in which IR spectroscopy, KF titration and vacuum drying are compared [6707]. One can expect that a good grinding, together with a titration at elevated temperatures and the application of pyridine-free reagents will result in an improvement of the water determination in vegetables.

Dried fruit also consists of mainly intact plant tissue which contains enclosed cell juices. The samples have to be homogenized first and then further triturated by "high-frequency shredding" and subsequently titrated at room temperature. A collaborative study on sultanas [8003] gave acceptable results.

9.6.15 Meat and Meat Products

Zürcher and Hadorn [8101] also employed the "high-frequency shredding" with these products. The water content was then titrated at room temperature.

10 Technical Products, Natural Products

10.1 Technical Gases

The fundamentals of the water determination in gases are summarized in Section 6.6.5; several specific techniques are to be mentioned here.

The water contained in gases must be transferred into the liquid phase in order to make it accessible for the KF titration. By means of a special apparatus, Muroi [6101] passed oxygen, nitrogen and hydrogen through methanol. Then he titrated the absorbed water with a diluted KF reagent which had a WE of 0.5–1.0 mg/ml. This principle has since been modified in different ways. Archer and Hilton [7405] use the KF reagent as the absorption agent. Davies [7502] uses ethylene glycol and pyridine (4 + 1) for the absorption, adds an excess of KF solution and passes the gas through this mixture until the KF reagent has been consumed. For the investigation of natural gas, Below [7503] used a big flask (4 l volume) both as titration vessel as well as for measuring the volume of the gas. The flask is filled with 200 ml methanol, evacuated and dehydrated by pre-titration. The gas is then allowed to enter. After a period of time, the water is absorbed in the methanol and can be titrated.

Sistig [7103] recommends a continuous coulometric water determination for process control of gases in polymer chemistry (see Section 6.3).

Water can be frozen out of those gases which do not condense. Archer and Hilton [7405] condense the water at low temperature (dry ice/acetone) in a column. The water is then desorbed and carried into the KF cell by means of a stream of nitrogen. Using this enrichment technique, water contents as low as 0.2 ppm can be analysed. Muroi and Ono [7301] isolated small amounts of water out of gaseous hydrogen chloride in a similar way (refer to Section 8.1).

For liquefied gases, the liquid phase can also be incorporated into the analytical procedure. The gas is first condensed in deep-cooled methanol and then evaporated by warming. The water remains in the methanol and can be titrated. Muroi, Ono and Ishii [6801] extracted the water from the liquefied sample with the anhydrous solvent in a special pressure cylinder.

The determination of water in aerosol propellants and refrigerants is described in ASTM E 700–79 [7907]. Methanol is used as a scrubber liquid. It is pre-titrated in a 500 ml round-bottomed flask. Then the gas is transferred slowly through the scrubber liquid and the absorbed water is titrated with the KF reagent. The handling of gaseous and liquefied samples is described in detail.

Coulometry is particularly suited to the determination of water in gases. As it is a micro-method, it naturally allows a simple working technique (see Section 3.3). It is therefore to be expected that KF coulometry will soon dominate in the field of gas analysis.

10.2 Liquid Products of Mineral Oil

Mineral oil products – mainly hydrocarbons – are only soluble in methanol to a limited extent. They also have a limited capacity to dissolve water, i.e. the water contents to be determined are usually less than 100 ppm.

This requires particularly careful sample handling. Bottles used for storing samples should be dried in a drying cabinet and allowed to cool in a dessicator, as moisture can otherwise infiltrate. On the other hand, these products sometimes contain dispersed water or are saturated with water at higher temperatures, which can then separate out onto the walls of the bottle during storage. Addition of small amounts of an anhydrous alcohol like 1% isopropanol can inhibit this separation. If the samples are administered with a syringe, the syringe should first be rinsed 2 or 3 times with the sample.

A titration in methanol can result in the formation of two distinct phases. The KF reaction is not directly affected by this as the methanol phase extracts the water out of the oil phase during the course of the titration. This can lead to indication problems, however (see Section 7.1). A titration in a homogeneous phase is better. 1-Propanol, a mixture of 1-propanol and methanol or even better a mixture of chloroform and methanol (2 + 1) are examples of suitable titration media for this and are described in ASTM D 1744 [7809] for example. Titrating crude oil, a mixture of toluene, chloroform and methanol (2 + 1 + 1) is to be applied as the working medium.

Using a two-component reagent, a mixture of toluene, chloroform and solvent component is to be preferred.

The extraction of mineral-oil fractions with ethylene glycol and the subsequent titration of the ethylene glycol is also a possibility. A high enrichment of the water can thereby be achieved. This working method, however, should remain just as much an exception as azeotropic distillation with benzene or pyridine.

Due to the low water contents, a coulometric determination proves particularly practical. It enables a quick and reliable determination. In order to improve the solubility, chloroform can be added to the anolyte, although only in limited amounts as the conductivity will otherwise be lowered too much. Coulometry is not to be recommended in the case of crude oil, as it contains insoluble impurities.

Chemical effects are caused by mercaptans which are oxidized by iodine (see Section 7.8). Tetraethyl lead also reacts with iodine as in Eq. (10.1) thereby simulating a water content.

$$Pb(C_2H_5)_4 + I_2 \rightarrow Pb(C_2H_5)_3I + C_2H_5I. \tag{10.1}$$

There are a number of national and international standards dealing with the investigation of mineral-oil products. ASTM D 1744–64 [7809] recommends a methanolic one-component reagent with a WE of 2–3 mg/ml. 50 ml of methanol/chloroform (1 + 3) are pre-dehydrated, 50 ml of sample added, and then titrated. ISO/DIS 6296 [8205] recommends commercial KF reagents and 2-methoxyethanol as the preferred working medium. As well as these solvents, DIN 51 777 part 1 [8303] also recommends methanol/propanol mixtures (1 + 1) as the working medium and also refers to the classic as well as to the new pyridine-free KF reagents.

10.3 Insulating Oils, Lubricating Oils and Greases

These mineral-oil products have high boiling points, are usually very viscous and have only a slight capacity to dissolve water, and are insoluble or only very slightly soluble in methanol. Kawinski therefore titrated the water content in inhomogeneous phases [6102].

If a homogeneous working solution is to be achieved, the methanol must be partially replaced by chloroform. In ASTM D 1533–79 [7908], 100 ml of a chloroform/methanol mixture (2 + 1) is used for the titration of 25–50 ml samples of insulating oil. Coulometry is recommended as an alternative method, as the water contents are very slow (less than 75 ppm). Anolytes containing chloroform according to Bonderovskaja, Kropotova and Maško [7501] also are to be preferred in this case.

DIN 51 777 part 2 [7407] recommends the evaporation technique for mineral-oil products which have a boiling point of 200 °C and above. The sample is heated to 120 °C and the water thus evaporated is carried by a stream of nitrogen into pre-dehydrated methanol. For substances affected by heating the water is evaporated under vacuum at 60 °C and condensed into methanol at −79 °C.

Muroi, Ogawa and Ishii [6604] employed a similar technique for lubrication greases. The grease is dissolved in hot, dehydrated insulating oil and the water is evaporated and carried into the titration cell by a stream of dry nitrogen.

10.4 Plastics

Plastics bind water in their structure sometimes very strongly so that it diffuses out only very slowly on drying. The KF determination of water is therefore as uncertain as a loss on drying. Three different techniques of KF determination are employed: titration of the suspended product, titration of the dissolved product, and the thermal evaporation of the water and its subsequent titration.

As a method of production control, Glöckner and Meyer [5903] titrated the water content by extraction of polyamide chips with methanol for 60 min at room temperature, or for 20 min at 35 °C. The reproducibility is much better than with oven drying. DIN 53 715 [7306] covers the analysis of plastics in powdered form (particle size less than 0.2 mm) by a similar method although without giving an extraction time.

Reinisch and Dietrich [6708] found that the water in polycaprolactam does not completely diffuse out. They therefore dissolved the samples in methanol/phenol (1 + 2) and titrated the total water content. Kellum and Barger [7005] recommend trifluoroethanol or hexafluoroisopropanol as solvents for high-molecular nylon. The dissolution takes 1–3 h. Other solvents can also be used to dissolve plastics, for example chloroform for polymethacrylates and formaldehyde/urea condensates. The plastic may precipitate when the solution is introduced into the alcoholic working medium necessary for the KF titration.

Muroi [6202] employed an evaporation technique for the investigation of polyethyleneterephthalate, nylon, polyethylene, polypropylene and polycarbonate. The plastic chips are heated to 100–200 °C and the water thus evaporated is carried by a nitrogen stream into a mixture of methanol and ethylene glycol (1 + 1). Reid and Turner [6103] analysed polyethylene in the same manner. Dinse and Praeger used the same working technique for the investigation of polyethyleneterephthalate [6908] and polyamide [7007]. ASTM D 789–81 [8110] prescribes a modified procedure for the investigation of nylon in which the water is evaporated under vacuum and then condensed at −70 °C.

10.5 Ion Exchange Resins

These products are cross-linked polystyrene resins which are insoluble in organic solvents. They can therefore only be titrated in suspension. Pollio [6307] used either pyridine or methanol as the working solvent for the investigation of a strongly acidic macro-reticular cation exchanger. Heumann and Rochon [6605] attempted to apply this procedure to Dowex 50W–X8. They found titration times of up to 2 h and water contents which were too low. Sharma and Subramanian [6907] investigated different Dowex 50 resins and found good agreement with the oven-drying method. The same working technique is also suitable for anion exchangers [7006]. Van Acker, de Corte and Hoste [7404] pointed out a possible iodine addition to the anion exchanger. Evenko, Kats and Sherman [7803] titrated weakly basic anion exchangers in the same way.

10.6 Cellulose, Paper, Wood

Cellulose and wood appear to release water easily and can therefore be easily titrated. Mitchell [4002] left cotton fibres to stand for an hour in methanol and then titrated the suspension. As usually, the water content found here was slightly higher than by drying, as a certain residual moisture always remains with the drying-cabinet method. Newspaper, cardboard and wood shavings were investigated in the same way.

Wood (cypress, douglas fir, oak) was tested using 3 mm thick boards. The samples (1 × 0.75 × 0.125 in.) were extracted twice in succession, for 2 h initially and then for 18 hours. Slightly higher values than by oven drying at 102 °C were given here as well, 7.96% as opposed to 7.69% for cypress wood for example.

ASTM D 1348–61 [7906] recommends a KF titration as well as an oven-drying method for the determination of moisture in cellulose whereby both a direct titration as well as a back titration can be employed.

10.7 Insulating Paper

A low water content is an important requirement for the suitability of these papers so that the KF titration can also be applied here. Gimesi, Cserfalvi and Pungor [7305] extracted the samples with methanol and titrated the water contents. ASTM D 3277–73 [7811] modifies the working conditions in that a methanol/chloroform mixture (2 + 1) is used for the extraction. Fujino, Muroi and Morishita [8108] compared the results of all the methods. Evaporating the water at temperatures of between 105 °C and 130 °C and transferring it by a stream of nitrogen into a KF fixture gave the most accurate results.

10.8 Surfactants

The water content of these products in powder, paste or liquid form can be titrated directly as described in ISO 4317 [7710] whereby a two-component technique can also be employed.

10.9 Cigarette Smoke Condensate

In order to determine the dry, nicotine-free smoke condensate the cigarettes are mechanically "smoked". The raw condensation precipitates out on cooling. The water content thereof is titrated according to DIN 10 240 part 3 using a one-component reagent. After determining the alkaloid content, the dry, nicotine-free smoke condensate is then calculated [7807].

10.10 Paints and Varnishes

ASTM D 1364–78 [7808] describes the determination of water in solvents for the paint industry. A pyridine/ethylene glycol mixture is used as the working medium. For the investigation of ketones and organic acids, it is recommended to cool with ice. The determination of water in terpentine, pinene, dipentene and other oils is described in ASTM D 890–58 [8109]. The two-component technique is used; the solvent component contains benzene. ASTM D 4017–81 [8113] recommends pyridine as the working medium and a one-component reagent as the titration agent for the investigation of all of the raw materials including the plastic resins.

10.11 Fertilizers

A determination of the adherent moisture is of significance as this is responsible for a hardening of the product during storage. Titrating the adherent moisture, the hydrate water will also be determined at the same time as it is only weakly bound. A direct KF titration is therefore not possible. The adherent moisture must be dissolved and separated by a solvent. Johnston and Smith tried out different mixtures of methanol and chloroform [6404]. Caro [6405] preferred to use dioxane as the separating agent and the values thus obtained proved the most reliable for various fertilizers. Duncan and Brabson [6909] checked this working method and confirmed that it was indeed only adherent moisture that was determined. The total water content can be determined by azeotropic distillation using n-amyl alcohol followed by KF titration.

10.12 Cement

A direct titration of cement is not possible as, on the one hand, the water is chemically bound, and on the other, the oxides present can form water by reaction with the KF reagent.

According to Rechenberg [7604], the water can be thermally driven off at 1000 °C in a tube furnace, then passed over an oxidation catalyst and absorbed and titrated in a KF cell. This has several advantages over the cement standard DIN 1164 part 3 (calcination). It determines the actual water content and not the carbon dioxide which is equally driven off and which is also determined by the calcination method. The errors inherent in calcining due to reducing substances [sulphides, metallic iron, ferrous oxide, manganese(II)oxide] are eliminated by the oxidation catalyst and do not influence the KF titration.

10.13 Minerals

A determination of the water content by calcining is uncertain as the carbonates present lead to higher values and the oxidizable substances combine with oxygen and negatively put the result in error. A direct titration is equally non-applicable. It would determine the adherent moisture but could be put in error by soluble oxides and carbonates. For these reasons, the gaseous mixture liberated by calcining, consisting mainly of water and carbon dioxide, is passed over to a KF titration. Hibbits and Zucker [5401] analysed uranium oxide by passing the desorbed moisture into the cell containing a KF solution. Lindner and Rudert [6905] described in detail an apparatus which can be used for the investigation of various minerals. Jecko and Ravaine [7504] used the same technique for the investigation of iron ores. Farzaneh and Troll [7701] optimized the working conditions in order to reduce the times taken for the analysis and also determined the standard deviation of investigations on various minerals. Yamaya [7902] applied the working method to the investigaton of the high fluoride-containing muscovite.

Literature

Monographs

Eb.   Eberius, E.: Wasserbestimmung mit Karl-Fischer-Lösung, 2nd Ed. 1958, Verlag Chemie Weinheim.

Aqu.   Mitchell jr., J.; Smith, D.M.: Aquametry, Part III (The Karl Fischer Reagent), 2nd Ed. 1980, John Wiley & Sons.

Hyd.   Hydranal – Pyridine-free Karl Fischer Reagents, Booklet of Riedel-de Haën AG, D-3016 Seelze-Hannover, West Germany. 3rd English Ed. 1982.

Publications

3501   Fischer, K.: New Method for the Volumetric Determination of Water Content in Liquids and Solids. (In German.) Angew. Chem. 48, 394 (1935).
Original publication of the KF method. Recommended reagent: pyridine 1.7 mol/l, iodine 0.33 mol/l, sulphur dioxide 0.5 ml/l in methanol. Working instructions for liquid and solid substances.

3901   Smith, D.M.; Bryant, W.M.D.; Mitchell, J.: Analytical Procedures Employing Karl Fischer Reagent. I. Nature of the Reagent. J. Am. Chem. Soc. 61, 2407–2412 (1939).
The stoichiometric course of the KF reaction is determined and a new reaction equation is given. Studies concerning side reactions and different compositions of the KF solution.

4001   Bryant, W.M.D.; Mitchell, J.; Smith, D.M.: Analytical Procedures Employing Karl Fischer Reagent. V. The Determination of Water in the Presence of Carbonyl Compounds. J. Am. Chem. Soc. 62, 3504–3505 (1940).
Aldehyde and ketone interference in the Karl Fischer titration for water is inhibited by reaction with an excess of hydrogen cyanide. The resulting cyanhydrines are inert toward the reagent.

4002   Mitchell, J.: Determination of Moisture in Native and Processed Cellulose. Ind. Eng. Chem. Anal. Ed. 12, 390–391 (1940).
Moisture in various forms of cellulose is determined by extraction with methanol followed by titration. Recovery is more nearly complete than by oven drying.

4101   Bryant, W.M.D.; Mitchell, J.; Smith, D.M.; Ashby, E.C.: Analytical Procedures Employing Karl Fischer Reagent. VIII. The Determination of Water of Hydration in Salts. J. Am. Chem. Soc. *63*, 2924–2927 (1941).

The determination of water in hydrated salts has been demonstrated by analyses of 25 typical hydrates. The water content of inorganic desiccants can be determined.

4102 Mitchell, J.; Smith, D.M.; Ashby, E.C.; Bryant, W.M.D.: Analytical Procedures Employing Karl Fischer Reagent. IX. Reactions with Inorganic Oxides and Related Compounds. Oxidation and Reduction Reactions. J. Am. Chem. Soc. *63*, 2927–2930 (1941).

Oxides and hydroxides react with the KF solution producing water. Carbonates and sulphites give the same reaction. Thiosulphate, sulphides and nitrites reduce the iodine. Cu(II) and As(V) oxidize the iodide.

4301 Wernimont, G.; Hopkinson, F.J.: The Dead-Stop End Point as Applied to the Karl Fischer Method for Determining Moisture. Ind. Eng. Chem. Anal. Ed. *15*, 272–274 (1943).

The dead-stop method gives a sharp and reproducible end point in nonaqueous solutions. An excess of KF reagent is added to the sample, the excess is titrated with water containing methanol. Water determination in methanol, aceton and solvent mixtures containing methanol.

4501 Johnson, C.M.: Determination of Water in Dry Food Materials. Application of the Fischer Volumetric Method. Ind. Eng. Chem. Anal. Ed. *17*, 312–316 (1945).

Determination of moisture in starches, sugars, pectin, spray-dried egg powder and dehydrated vegetables. The results are compared with those obtained by the standard vacuum-oven technique.

4701 Johansson, A.: Determination of Water by Titration. A Modified Karl Fischer Method. Svensk. papperstidning, Stockholm, *50*, 124–126 (1947).

The KF reagent is applied in two separate solutions. The solution A consisting of sulphur dioxide, pyridine and methanol is used as a solvent for the sample. A methanolic solution of iodine serves as a titrant.

4702 di Caprio, B.R.: Rapid Determination of Moisture in Liquid Sulfur Dioxide. Anal. Chem. *19*, 1010–1011 (1947).

30 ml of sulphur dioxide are taken into 30 ml of methanol and titrated with the KF reagent. Sulphur dioxide is hygroscopic and needs a special handling.

4801 Brobst, K.M.: Determination of Moisture in Lecithin and Crude Soybean Oils. Anal. Chem. *20*, 939–941 (1948).

A mixture of chloroform and methanol (3 + 1) is pre-titrated with the KF reagent. Then the sample and an excess of KF reagent are added. Back-titration with water-in-methanol solution.

4802 Schroeder, C.W.; Nair, J.H.: Determination of Water in Dry Food Materials. Anal. Chem. *20*, 452–455 (1948).

The samples are extracted refluxing with methanol for 15 min or at room temperature for 2 h and more. The KF method offers certain advantages over the more commonly used oven-drying methods.

4902 Seaman, W.M; McComas, W.H.; Allen, G.A.: Determination of Water by Karl Fischer Reagent. Stoichiometric Iodometric Method. Anal. Chem. *21*, 510–512 (1949).

According to Johansson two seperated solutions are used. Both solutions are stable in storage. The reaction is stoichiometrical according to the known equation.

4903 Milberger, E.C.; Uhrig, K.; Becker, H.C.; Levin, H.: Determination of Water in Hydrogen Chloride by Means of Karl Fischer Reagent. Anal. Chem. *21*, 1192–1194 (1949).
Hydrogen chloride is absorbed in a cooled mixture of methanol, pyridine and KF reagent. Back titration with water-in-methanol.

5101 van der Meulen, J.H.: The Determination of Water according to Karl Fischer and the Analysis of the Reagent. (In Dutch.) Chem. Weekbl. Mag. *47*, 608–609 (1951).
Description of the KF method. The content of active iodine is determined by two titrations of the content of sulphur dioxide. Before starting the second titration iodine is absorbed using mercury.

5102 Linch, A.L.: Determination of Purity and Water Content of Xanthates and Dithiocarbamates. Anal. Chem. *23*, 293–296 (1951).
Direct titration of water in chloroform using a methanolic KF reagent. The active thiol group does not react with the iodine in the KF solution.

5201 Aktiebolaget Pharmacia, Upsala / Blomgren, E.; Jenner, H.: Improvements in and relating to quantitative determination of water. BP 722, 983.
The Karl Fischer reagent is stabilized by adding iodide. In such a solution the titer does not change with time.

5202 Laurene, A.H.: Determination of Water by Karl Fischer Titration in the Presence of Ferric Salts. Anal. Chem. *24*, 1496 (1952).
Ferric salts interfere in the KF titration liberating iodine. The oxidation of the iodide may be eliminated by complexing with 8-hydroxy-quinoline.

5203 Hodgson, H.W.; Glover, J.H.: The Determination of Water in Liquid Ammonia. Analyst (London) *77*, 74–77 (1952).
The ammonia is removed by evaporation after addition of ethylene glycol to retain the water. The water is titrated with Fischer reagent after neutralization of traces of residual ammonia with acetic acid.

5301 van der Meulen, J.H.: Reagent for the Quantitative Determination of very Small Amounts of Water in Gases, Liquids and Solid Substances, and Method of Preparing the Reagent. BP 728, 947.
The previously used pyridine is replaced by acetates, preferably sodium acetate. Other acetates or salts of weak carboxylic acids may be used as well.

5302 Blasius, E.; Wachtel, F.: Investigations to Detect a Iodine-Pyridinium-Complex Using Ion-Exchange. (In German.) Fresenius Z. Anal. Chem. *138*, 106–110 (1953).
Methanolic solutions containing iodine and pyridine produce only small amounts of the (Pyr.I)-cation.

5401 Hibbits, J.O.; Zucker, D.: Determination of Absorbed Moisture on Uranium and Uranium Oxide. Anal. Chem. *26*, 1093–1095 (1954).
The sample was placed in a quartz tube and heated, the evolved moisture swept into a titration cell where it reacted with the reagent.

5501 Peters, E.D.; Jungnickel, J.L.: Improvements in Karl Fischer Method for Determination of Water. Anal. Chem. *27*, 450–453 (1955).

A substantial gain in stability is achieved by substituting 2-methoxyethanol for methanol in the KF reagent. A mixture of ethylene glycol and pyridine permits the direct titration of water in ketones.

5502 Bonauguri, E.; Seniga, G.: The Mechanism of the Karl Fischer Titration and a new Modification of this Method. (In German.) Fresenius Z. Anal. Chem. *144*, 161–164 (1955).
Titrating water the reagent consumption depends on the water-to-methanol ratio in the solution. A KF reagent containing dioxane is recommended.

5601 Johansson, A.: Application of Karl Fischer Water Method to Oxidants, Reductants, and Amines. Anal. Chem. *28*, 1166–1168 (1956).
The two-solution modification has special advantages with reducing and oxidizing substances and amines. Substitutes for pyridine and methanol were tried.

5602 Moberg, M. L.; Knight, W. P.; Kindsvater, H. M.: Determination of Water Content of White Fuming Nitric Acid Utilizing Karl Fischer Reagent. Anal. Chem. *28*, 412–413 (1956).
The sample is neutralized by the use of pyridine-dimethylformamide solution. An excess of KF reagent is then added and back-titrated with a water-methanol solution. Nitrogen dioxide concentrations of less than 1.5% do not interfere.

5603 Eberius, E.; Kowalski, W.: The Effect of the Content of Methanol on the Determination of Water using the Karl-Fischer-Solution. (In German.) Fresenius Z. Anal. Chem. *150*, 13–20 (1956).
Determinating the titre of the KF solution, the amount of methanol used effects the consumption of KF solution and the calculated titre.

5701 McComb, E. A.: Formamide as an Extraction Solvent in Karl Fischer Method for Determining Moisture in Lactose and Maltose. Anal. Chem. *29*, 1375 (1957).
Lactose and maltose are soluble in formamide and the water content can be titrated using this solvent.

5702 Aktiebolaget Pharmacia, Upsala / Blomgren, E.; Jenner, H.: Stabilized Karl Fischer Reagent. DBP 1 075 341, USP 2,967,155.
The KF reagent is stabilized by adding iodide ions and an amine from the group of secondary and tertiary aliphatic amines. The optimum concentrations of iodide and amine can be calculated.

5703 Eberius, E.; Kempf, W.: The Determination of Water in Starch, Starch Derivates and Byproducts using the Karl Fischer Method. (In German.) Staerke *9*, 77–81 (1957).
Starch products are extracted with methanol and the water content is titrated with KF solution. The determination is quick, accurate and the results are highly reproducible. The results are about 0.5% higher than those obtained by drying in a Brabander moisture tester.

5801 Oehme, F.: Procedure for the Coulometric Determination of Water. (In German.) DBP 1 086 918.
Karl Fischer solution containing alkyliodide instead of iodine.

5802 Seibel, W.: Method for the Volumetric Determination of Water in Grain using Karl Fischer Solution. (In German.) Getreide und Mehl *8*, 1–4 (1958).
The samples are grinded, suspended in methanol and titrated with KF solution at 50 °C. The titration takes about 30 min.

5803 Seibel, W.; Bolling, H.: Automatic Determination of Water in Grain Products using Karl Fischer Solution. (In German.) Getreide und Mehl *8*, 73–76 (1958).
Using the method of [5802] wheat, rye, barley, corn, rice, flour, farinaceous products and bread are analyzed. Comparison of the results with oven-drying.

5901 Meyer, A.S.; Boyd, C.M.: Determination of Water by Titration with Coulometrically Generated Karl Fischer Reagent. Anal. Chem. *31*, 215–219 (1959).
An automatic titrator and a reagent are described. The water content in liquid products can be easily and reliably determined. Diamines need a neutralization with salicylic acid and ethylene glycol as a solvent.

5902 Barnes, L.; Pawlak, M.S.: Direct Karl Fischer Determination of Water in Vinyl Ethers. Anal. Chem. *31*, 1875–1876 (1959).
The interferences caused by the vinyl ethers are eliminated by the use of a stabilized reagent and a pyridine-acetic acid solvent.

5903 Glöckner, G.; Meyer, W.: The Determination of the Water Content in Polyamide Chips using Karl Fischer Reagent and the Dead-Stop Indication. (In German.) Faserforsch. Textiltech. *10*, 83–84 (1959).
The chips are extracted with methanol, either 60 min at room temperature or 20 min at 35 °C. Then the water is titrated with KF solution.

6001 Sandell, D.: Application of the Karl Fischer Method to the Determination of Water in Sugar Confectionary Materials. J. Sci. Food Agric. *11*, 671–678 (1960).
About 30 different sugars and carbohydrates are analyzed. The samples are extracted with methanol and titrated with KF solution. Comparable results obtained by KF and oven-drying are presented.

6101 Muroi, K.: Determination of a Micro Quantity of Water in Gaseous Samples by the Karl Fischer method. (In Japanese.) Bunseki Kagaku *10*, 841 (1961) (Engl. summary).
Gaseous samples, such as oxygen, nitrogen or hydrocarbons, are passed through anhydrous methanol. The water is completely absorbed and can be titrated using a diluted KF reagent.

6102 Kawinski, H.: The Determination of Small Amounts of Water in Insulating Oils. (In German.) Erdöl, Kohle, Erdgas, Petrochem. *14*, 271–275 (1961).
50 ml of the sample are added to 100 ml of pre-dried methanol. The sample is not soluble in the solvent but the water content can be titrated completely.

6103 Reid, V.W.; Turner, L.: The Determination of Water in Plastic Materials. Analyst (London) *86*, 36–39 (1961).
A stream of dry nitrogen passes over the sample, which is maintained at 120 °C. The water evaporated is carried by the nitrogen into a KF cell. The determination takes about 30 min.

6201 Burns, E.A.; Muraca, R.F.: Karl Fischer Determination of Water in Ammonium Perchlorate with Automatic Titration Apparatus. Anal. Chem. *34*, 848–854 (1962).
Total water content is analyzed by dissolving the sample in pyridine-methanol (3 + 1). Surface water is titrated using a saturated solution of ammonium perchlorate as a working medium.

6202 Muroi, K.: Determination of Water in Plastic Materials by the Karl Fischer Method. (In Japanese.) Bunseki Kagaku *11*, 351 (1962) (Engl. summary).
Plastic chips are heated up to 200 °C. The vaporized water is carried with a stream of nitrogen into a mixture of methanol-ethylene glycol (1:1). Water determination in polyethylene terephthalate, Nylon, polycarbonate, polypropylene, polyethylene.

6301 Swensen, R. F.; Keyworth, D. A.: Determination of Water below 10 ppm in Benzene and Related Solvents, using Coulometrically Generated Iodine. Anal. Chem. *35*, 863–867 (1963).
Iodine is generated from potassium iodide dissolved in a solution containing pyridine, formamide and sulphur dioxide. A special cell and a system of sample handling prevent contamination by external moisture.

6302 Muroi, K.; Ogawa, K.: The Determination of the Water Content of the Low Polymeres of Formaldehyde by the KF Method. I. Absorbed Water and Bound Water Formed by Pyrolysis of the Polymeres. Bull. Chem. Soc. Jpn. *36*, 965–969 (1963) (Engl.).
Absorbed water is extracted using dry methanol. After 10 min the water is titrated with KF solution. To determine the total water content, the substance is dissolved at 160 °C in propylene glycol.

6303 Muroi K.; Ogawa, K.: The Determination of Water Content in Low Polymers of Formaldehyde by the KF Method. II. Total Amount of Absorbed and Bound Water by the Methanol Absorption Method. Bull. Chem. Soc. Jpn. *36*, 1273–1280 (1963) (Engl.).
The sample is pyrolyzed in a stream of dry nitrogen at 200–250 °C. The water formed is absorbed in methanol and titrated with KF solution.

6304 Chiang, P. T.; Cooper, W. C.: Determination of Water and Alkali Metal Oxide in Alkali Metal Salts of Selenium and Tellurium by Karl Fischer Titration. Anal. Chem. *35*, 1693–1695 (1963).
The water content can be determined. Selenite, tellurite and tellurate behave like a mixture of the respective oxides, and the alkali oxide reacts quantitatively with the KF reagent. Selenate does not react.

6305 Barbi, G. B.; Pizzini, S.: Determination of Water Content in Potassium Difluoride. Anal. Chem. *35*, 409–410 (1963).
The sample is dissolved in acetic acid and titrated after adding to pre-dried pyridine.

6306 Burns, E. A.; Muraca, R. F.: Determination of Water in Red Fuming Nitric Acid by Karl Fischer Titration. Anal. Chem. *35*, 1967–1970 (1963).
Nitrogen oxides interfere with the KF reaction. The sample is diluted with ethylene dichloride and neutralized with pyridine. Corrections for the content of nitrogen dioxide.

6307 Pollio, F. X.: Determination of Moisture in Ion Exchange Resins by Karl Fischer Reagent. Anal. Chem. *35*, 2164–2165 (1963).
Direct titration of the resin in either methanol or pyridine solvent.

6401 Fischer, F.; Schiene, R.: Determination of Small Amounts of Water in Aliphatic Ketones with Karl Fischer Solution using the Dead-Stop Method. (In German.) Z. Chem. (Leipzig) *4*, 69–70 (1964).
Pyridine is pre-dried by titration with the KF solution. The sample (acetone) is

added and titrated with the same KF solution, produced with 2-methoxyethanol as a solvent.

6402 Giuffre, L.: Determination of Small Amount of Water in Halogenides of Transition Metals. (In Italian.) Chim. Ind. (Milan) *46*, 51–53 (1964).

Titanium(III)-chloride and vanadium(III)-chloride are dissolved in benzene-ethanol mixtures. The water is separated by distillation and then titrated using a Karl Fischer solution.

6403 Thung, S. B.: Comparative Moisture Determination in Dried Vegetables by Drying after Lyophilisation or by the Karl Fischer Method. J. Sci. Food Agric. *15*, 236–244 (1964).

Drying was conducted in a vacuum desiccator at different temperatures. The results found at 50 °C are in good agreement with the values found by KF titration after extraction with formamide.

6404 Johnston, I.; Smith, E. J.: The Determination of Water in Fertilizers by a Modified Karl Fischer Method. Chem. Ind. (London) *8*, 315 (1964).

Free water is extracted with methanol chloroform mixtures and titrated with KF solutions. Various mixtures of chloroform and methanol are employed.

6405 Caro, J. H.: Free Water in Fertilizers by Karl Fischer Titration. J. Assoc. Off. Anal. Chem. *47*, 626–629 (1964).

The method involves extraction with dioxane and titration of the extract with KF solution. Preliminary tests showed dioxane to be the most feasible extractant for fertilizers.

6501 Archer, E. E.; Jeater, H. W.: The Determination of Small Amounts of Water in Some Organic Liquids. Analyst (London) *90*, 351–355 (1965).

N-Ethylpiperidine is used as a catalyst to increase the rate of reaction. End points are detected using polarized electrodes and a pH-meter. The cell is made over-dry before the titration is started.

6502 Muroi, K.; Ogawa, K.; Ishii, Y.: The Determination of the Water Content in Acetaldehyde by Means of the Karl Fischer Reagent. Bull. Chem. Soc. Jpn. *38*, 1176–1181 (1965).

The sample of acetaldehyde is added to a mixture of pyridine and propylene glycol and evaporated with dry nitrogen. The remaining water is titrated.

6503 Majewska, J.; Ogorzalek, A.: Some Methods for the Determination of Water in Polyamide by Karl Fischer Titration. (In Polish.) Chem. Anal. (Warsaw) *10*, 491–494 (1965).

The sample is dissolved in a mixture of phenol and trichloroethane at 45–50 °C and titrated with a KF solution.

6601 Beyer, H.; Varga, K.: The Determination of Water Content in Carbonyl Compounds using Karl Fischer Solution. Z. Chem. (Leipzig) *6*, 470–471 (1966).

Acetone, cyclohexanone and butyraldehyde are cooled to temperatures below −40 °C and titrated at low temperatures using pyridine as a solvent.

6602 Epps jr., E. A.: Determination of Water in Molasses by the Karl Fischer Method. J. Assoc. Off. Anal. Chem. *49*, 551–554 (1966).

Collaborative studies were conducted in 3 years to compare the KF method with the vacuum drying method. Good agreement with known water contents.

6603 Smith, R.C.; Kellum, G.E.: Modified Karl Fischer Titration Method for Determination of Water in Presence of Silanol and Other Interfering Materials. Anal. Chem. *38*, 67–72 (1966).

High molecular weight alcohols are used as working solution to minimize the interfering reaction which produces water. The method is also applicable to ketones and aldehydes.

6604 Muroi, K.; Ogawa, K.; Ishii, Y.: The Determination of the Small Amount of Water in Grease using the Karl Fischer Method. Bull. Jap. Petr. Inst. *8*, 45–50 (1966).

Insulating oil and the sample are placed in a boiling flask and heated to 120–140 °C. Dry nitrogen is passed through the flask, carrying the moisture into a mixture of methanol and ethylene glycol.

6605 Heumann, W.R.; Rochon, F.D.: Determination of Water in Cation Exchange Resins by Karl Fischer Reagent. Anal. Chem. *38*, 638–639 (1966).

Dowex 50W-X8 in its sodium form is suspended in methanol and titrated. Up to 2 h were needed to reach a stable end point and the results are too low.

6606 International Standard FIL/IDF 23:1964: Determination of the Water Content of Butter Oil by the Karl Fischer Method. Milchwissenschaft *21*, 137–139 (1966).

The sample is dissolved in a mixture of chloroform and methanol (3 + 1) and titrated using a methanolic KF solution.

6701 Muroi, K.; Ogawa, K.; Ishii, Y.: Determination of Water in Substances Containing Active Chlorine by Karl Fischer Titration. Bunseki Kagaku *16*, 1061 (1967) (Engl. summary).

To suppress the interference, active chlorine is made inactive by pretreatment of the sample with a methanolic solution of pyridine and sulphur dioxide.

6702 Bizot, J.: Automatic Method for the Coulometric Determination of Small Amounts of Water. (In French.) Bull. Soc. Chim. France 1967, 151–157.

An apparatus and different cells for the coulometric determination of liquids, solids and gases are described. The KF reagent consists of formamide.

6703 Klimova, V.A.; Sherman, F.B.; L'vov, A.M.: Micro-Determination of Water with Fischer Reagent, Prepared in Dimethylformamide in Place of Methanol. Bull. Acad. Sci. (USSR) Div. Chem. Sci. (Engl.) 1967, 2477–2479.

A one-component reagent is prepared consisting of iodine, sulphur dioxide, pyridine and DMF as a solvent. Water determination in aldehydes and ketones, using DMF as a solvent for the sample.

6704 Klimova, V.A.; Sherman, F.B.; L'vov, A.M.: A General Titrimetric Micro Method for the Determination of Water by the Fischer Reagent. Bull. Acad. Sci. (USSR) Div. Chem. Sci. (Engl.) 1967, 2631–2633.

The titer of the reagent described in [6703] depends on the solvent used for the sample. The titre is decreased by the factor of two if methanol is used instead of DMF. Two different equations are given for the KF reaction.

6705 Kellum, G.E.; Smith, R.C.: Determination of Water in Monomeric and Short-Chain Silanoles Employing a Modified Karl Fischer Titration Method. Anal. Chem. *39*, 1877–1879 (1967).

Silanols react with the alcohol present in the KF reagent releasing water.

Short-chain silanols react more rapidly than long-chain ones and aromatic ones. KF titration at low temperatures and in higher alcohols as solvents.

6706 Kellum, G. E.; Smith, R. C.: Determination of Water, Silanol, and Strained Siloxane on Silica Surfaces. Anal. Chem. *39*, 341–345 (1967).
Absorbed moisture is titrated directly using a mixture of 2-ethylhexanol and pyridine as a solvent. Methanol reacts with strained siloxane groups producing water.

6707 Rader, B. R.: Determination of Moisture in Dried Vegetables. J. Assoc. Off. Anal. Chem. *50*, 701–703 (1967).
A collaborative study on moisture, in which a near-infrared spectrophotometric method and the KF method are compared with the vacuum oven method.

6708 Reinisch, G.; Dietrich, K.: The Determination of Water in Polycaprolactame. (In German.) Faserforsch. Textiltech. *18*, 535–536 (1967).
The sample is dissolved in a mixture of phenol and methanol (2 + 1) at 60 °C within 1–2 h. Then the water content is titrated.

6801 Muroi, K.; Ono, M.; Ishii, Y.: Rapid Determination of Water in Liquefied Gases by Karl Fischer Method. Bull. Jap. Petr. Inst. *11*, 34 (1968) (Engl. summary).
The KF method can be applied after absorption of water with an anhydrous solvent from the liquefied sample in a special pressure cylinder.

6802 della Monica, E. S.; Holden, T. F.: Comparison of Toluene Distillation and Karl Fischer Methods for Determining Moisture in Dry Whole Milk. J. Dairy Sci. *51*, 40–43 (1968).
Comparative study. The sample is extracted with methanol. An aliquote is used for the KF titration.

6901 Strange, T. E.: Water Content in Cheese: Comparison between the Distillation Method and the Karl Fischer Method. J. Assoc. Offic. Anal. Chem. *52*, 117–125 (1969).
Cheese is homogenized in n-butanol and the water is extracted. The water contents obtained by KF titrations agree with those from distillations.

6904 Seltzer, D. M.; Levy, G. B.: A Novel Micro Method of Water Determination. Am. Lab. (Fairfield, Conn.) 1969, Sept. 30–32.
A newly developed instrument for the coulometric water determination is presented: the Aquatest. The advantages of coulometry are mentioned.

6905 Lindner, B.; Rudert, V.: An Improved Method for the Determination of Bound Water in Rocks, Minerals and Other Solids. (In German.) Fresenius Z. Anal. Chem. *248*, 21–24 (1969).
For the determination of water the samples are decomposed by heating. A stream of dry carrier gas takes the liberated water into a Karl Fischer cell.

6906 Muroi, K.; Tsutsumi, C.; Koizumi, H.: Determination of Moisture in Caramel and Sugar Candy: Comparison of the Karl Fischer Method using Formamide as Solvent with a Film Method. J. Food. Sci. Technol. Jap. *16*, 39 (1969) (Engl. summary).
Caramel is dissolved in a mixture of formamide and methanol at 55–60 °C and titrated with KF solution. Sugar candy dissolves at room temperature.

6907 Sharma, H.D.; Subramanian, N.: Determination of Water in Ion-Exchange Resins by the Karl Fischer and Drying Methods. Anal. Chem. *41*, 2063–2064 (1969).
Direct titration with KF reagent gives rapid and accurate estimation of water content for the sulphonated cation exchange resins. For swollen resins the titration takes less than 10 min, dried resins need more time.

6908 Dinse, H.-D.; Praeger, K.: A new Method for the Determination of Water in Polyethylene Terephthalate. (In German.) Faserforsch. Textiltech. *20*, 449–450 (1969).
The sample is heated to 170 °C. The evaporated water is carried with a stream of dried nitrogen into a KF cell. The determination takes about 30 min.

6909 Duncan, R.D.; Brabson, J.A.: Determination of Free and Total Water in Fertilizers by Karl Fischer Titration. J. Assoc. Off. Anal. Chem. *52*, 1127–1131 (1969).
Free water is extracted with dioxane and titrated. Total water is determined by distillation of n-amyl alcohol from its mixture with the sample and titration of the distillate.

7001 Lugowska, M.; Hopfinger, A.; Fligier, J.; Czichon, P.: Determination of Water in Dodecyl Benzene Sulphuric Acid using the Fischer Method. (In Polish.) Chem. Anal. (Warsaw) *15*, 449–451 (1970).
The acid is neutralized by pyridine and the water content is determined by back titration. Sulphuric acid does not cause interferences.

7002 Sailly, M.; Belin, P.: Titer Determination of Karl Fischer Solutions. (In French.) C. R. Acad. Sci. (Paris) Ser. C *270*, 1710–1713 (1970).
Water is recommended as a primary standard for KF solutions. It is more accurate than solid hydrates or water-in-methanol mixtures.

7003 Muroi, K.; Ono, M.: Determination of Water in Strongly Basic Amines by the Karl Fischer Method. (In Japanese.) Bunseki Kagaku *19*, 1280–1282 (1970) (Engl. summary).
Strong basic amines can be applied to the KF titration in limited amounts. They can be titrated in a mixture of methanol and acetic acid (5 + 1).

7004 Delmonte, C.S.: Modified Karl Fischer Reagent. USP 3,656,907.
Reagent containing a sulphoxide or an organic nitrite as a reducing agent, a base which is not pyridine but containing the pyridine skeleton, iodine and methanol as a solvent.

7005 Kellum, G.E.; Barger, J.D.: Determination of Water in Nylon with Karl Fischer Reagent using Fluorinated Alcohols as Solvents. Anal. Chem. *42*, 1428–1429 (1970).
The samples are dissolved in hexafluoroisopropanol or trifluoroethanol (2–3 h) and then titrated with a KF solution.

7006 Sharma, H.D.; Subramanian, N.: Determination of Water in Ion-Exchange Resins: Anion Exchange Resins. Anal. Chem. *42*, 1287–1290 (1970).
The water content of Dowex AG1 and Bio-Rex 9 is titrated directly. The titration takes only 5 min. Comparison with drying methods.

7007 Praeger, K.; Dinse, H.-D.: Experiences in the Determination of Small Amounts of Water in Polyethylene Terephthalate and Polycapryle Amide. (In German.) Faserforsch. Textiltechn. *21*, 37–38 (1970).

The samples are heated to 170 °C. The evaporated water is carried into a KF cell and titrated. Comparison with drying methods.

7008 Strange, T.E.: Moisture in Cheese by Gas Chromatography and by Karl Fischer Titrimetry. J. Assoc. Off. Anal. Chem. *53*, 865–868 (1970).
15 g of cheese are mixed with 150 ml of propanol in a blender jar. A part of this solution is titrated with the KF reagent. Comparisons with GC.

7101 Muroi, K.; Ono, M.: Determination of Micro Amounts of Water in Organic Liquids by Karl Fischer Method. (In Japanese.) Bunseki Kagaku *20*, 975–979 (1971) (Engl. summary).
0.2–100 ppm of water are titrated using a KF reagent with a water equivalent of 0.1–0.5 mg/ml. To accelerate the titration, methanol with 8% of pyridine and 2.5% of sulphur dioxide is used as a working medium.

7102 Schneider, F.; Emmerich, A.; Ticmanis, U.: Determination of the Total Moisture Content in Crystallized White Sugar. (In German.) Zucker *24*, 245–249 (1971).
Sugar is dissolved in pre-dried formamide and titrated. The titration takes 1 h. To eliminate the 'drift' of the titration vessel, a blank titration is carried out.

7103 Sistig, E.: Continuous Determination of Traces of Water in Gas using a New Titration Cell. (In German.) Chem.-Ing.-Tech. *43*, 1061–1064 (1971).
The gas to be analysed is passed through the titration cell. The water contained is absorbed and titrated coulometrically.

7201 Andrysiak, A.; Andrysiak, E.: Potentiometric Determination of Water in Polyethers by the Karl Fischer Method. (In Polish.) Chem. Anal. (Warsaw) *17*, 1227–1231 (1972) (Engl. summary).
Polyethers produced from glycerol, pyrocatechol and phenol with ethylene oxide or propylene oxide are analysed. Disturbances by propion aldehyde.

7202 Thomasow, J.; Mrowetz, G.; Delfs, E.-M.: The Determination of Moisture in Dried Milk Products by Karl Fischer Titration. (In German.) Milchwirtschaft *27*, 76–81 (1972).
Water is titrated in milk powder and in whey powder. Comparison with oven drying gives lower values.

7203 Mika, V.; Čadersky, I.: Coulometric Determination of Microgram Amounts of Water in Silicone Compounds. Fresenius Z. Anal. Chem. *258*, 25–29 (1972).
Methanol reacts with silanol groups producing ethers. The effect of this side reaction was suppressed or eliminated by using 2-methoxyethanol in place of methanol.

7204 Photovolt Corporation, New York/Dahms, Levy, Seltzer. Method for Coulometric Karl Fischer Titration and Reagent Composition for Coulometric Karl Fischer Titration USP 3,682,783 and USP 3,749,659.
Reagent composition with 25–50% of sulphur dioxide, pyridine, iodide and methanol as a solvent.

7205 Meuser, F.; Kempf, W.: Methods for Determination of Water Content in Products from the Wet-Milling Process with Regard to their Influence on the Yield Calculation. (In German.) Stärke *24*, 167–171 (1972).
Water determination in corn, rice, wheat, starch and other products by KF titration, oven drying and GC. Extraction of the samples by methanol ore pyridine followed by KF titration.

7206 Strange, T.E.: Collaborative Study of Moisture in Cheese by Gas Chromatography and by Karl Fischer Titration. J. Assoc. Off. Anal. Chem. *55*, 507–510 (1972).
4 different samples are analysed in 5 laboratories using the methods described in [7008].

7301 Muroi, K.; Ono, M.: Determination of Trace Moisture in Hydrogen Chloride Gas by Karl Fischer Titration. Microchem. J. *18*, 234–239 (1973).
The moisture in the gas was collected in a cold trap at about $-80\,°C$ while the gas passed through, and then dissolved in pyridine-methanol (1:1).

7302 Sherman, F.B.; Zabokritskij, M.P.; Klimova, V.A.: Pyridine-less Fischer Reagent. Zh. Anal. Chim. (USSR) *28*, 1624–1625 (1973) (Engl. translation).
A procedure has been developed to prepare a modified Fischer reagent in which pyridine is replaced by sodium acetate.

7303 van Acker, P.; de Corte, F.; Hoste, J.: The Determination of Water in Copper(II)-Compounds by Karl Fischer Titration. Anal. Chim. Acta *67*, 236–239 (1973).
The sample solution is added to a methanolic solution containing an excess of KF reagent. Back-titration with methanol.

7304 Kviesitis, B.: Determination of Moisture in Molasses by the Karl Fischer Method. J. Assoc. Off. Anal. Chem. *56*, 1025 (1973).
The AOAC method has been modified to protect from atmospheric moisture.

7305 Gimesi, O.; Cserfalvi, T.; Pungor, E.: Application of the Karl Fischer Method to the Determination of the Moisture Content in Insulating Papers. Mikrochim. Acta 1973, 373–379.
The water content is extracted by methanol and titrated continuously with KF solution.

7306 DIN 53715. Determination of Water Content of Plastics in Form of a Powder by Titration According to Karl Fischer.
The powder (particle size less than 0.2 mm) is suspended in methanol and titrated using a diluted KF solution.

7401 Cedergren, A.: Coulometric Trace Determination of Water by using Karl Fischer Reagent and Potentiometric End-Point Detection. Talanta *21*, 367–375 (1974).
A specially constructed electrolysis cell was used. The reagent composition has been optimized in order to enhance the reaction rate. 0.05–200 microgram of water were analysed.

7402 Cedergren, A.: Reaction Rates between Water and the Karl Fischer Reagent. Talanta *21*, 265–271 (1974).
The reaction was found to be first-order with respect to iodine complex, to sulphur dioxide and to water. An excess of pyridine had no measurable effect on the reaction rate.

7403 Cedergren, A.: Comparison between Amperometric and True Potentiometric End-Point Detection in the Determination of Water by the Karl Fischer Method. Talanta *21*, 553–563 (1974).
Potentiometric end-point detection is more sensitive. Using this indication the titrations proceed faster.

7404 van Acker, P.; de Corte, F.; Hoste, J.: The Determination of Water in a

Strong Base Anion-Exchange Resin by the Karl Fischer Titration. Anal. Chim. Acta *73*, 198–203 (1974).

Dowex 1-X8 is extracted with a water-acetic acid mixture of known composition.

7405 Archer, E.E.; Hilton, J.: The Determination of Small Amounts of Water in Gases using Karl Fischer Reagent. Analyst *99*, 547–550 (1974).

The water is absorbed in the KF reagent or on a short chromatographic column at low temperatures.

7406 Miyake, S.; Suto, T.: An Automatic Coulometric Titrator for the Determination of Water by the Karl Fischer Method. Bunseki Kagaku *23*, 482–490 (1974).

The construction of the electric current source, of the counter circuit and the operation circuit are described. Less than 3 mg of water can be determined.

7407 DIN 51 777 part 2. Testing of Mineral Oil Hydrocarbons and Solvents. Determination of Water Content according to Karl Fischer Indirect Method.

Oils insoluble in methanol or 2-methoxyethanol are heated to 120 °C. The water is evaporated and carried into a KF cell using dried nitrogen.

7408 Miyake, S.; Suto,T.: Electrolytic Solution for Coulometric Generation of Karl Fischer Reagent. (In Japanese.) Bunseki Kagaku *23*, 476–482 (1974) (Engl. summary).

A stock solution consists of 0.35 mol/l potassium iodide, 7 mol/l sulphur dioxide and pyridine as a solvent. To get an anolyte it is diluted with 4 parts of methanol or an other solvent. The effect of the concentrations is investigated.

7501 Bondarevskaja, E.A.; Kropotova, E.D.; Masko, T.E.: Water Determination in Organic Liquids and Insulating Oils. (In Russian.) Zh. Anal. Chim. *30*, 560–564 (1975) (Engl. summary).

Coulometric titration using a new coulometric cell. The reagent contains chloroform to dissolve the sample.

7502 Davies, R.J.: The Determination of Water in Natural Gas using a Modified Karl Fischer Titration Apparatus. Analyst *100*, 163–167 (1975).

Sample gas is passed through an involatile solvent containing a measured volume of KF solution until the KF reagent is discoloured.

7503 Below, L.: Determination of Water in Gases according to Karl Fischer. (In German.) Gas und Wasserfach *116*, 480–482 (1975).

The water is absorbed in methanol and then titrated with a diluted KF reagent.

7504 Jecko, G.; Ravaine, D.: Rapid and Accurate Determination of Water in Iron-Ore using a Dynamic Karl Fischer Method. (In French.) Analusis *3*, 376–388 (1975).

The sample is heated to 900 °C and the evaporated water is carried into a titration cell using nitrogen gas.

7505 Schneider, F.; Emmerich, A.; Ticmanis, U.: Determination of Sugar Surface Moisture. (In German.) Zucker *28*, 349–356 (1975).

To titrate the adhering moisture the sample has to remain undissolved. Methanol saturated with sucrose is recommended as a suitable working medium.

7601 Bryan, W.P.; Rao, P.B.: Comparisons of Standards in the Karl Fischer Method for Water Determination. Anal. Chim. Acta *84*, 149–155 (1976).

Titration with continuous flushing of the titration cell. Sodium tartrate dihydrate is a good primary standard for KF reagents.

7602 Verhoef, J.C.; Barendrecht, E.: Mechanism and Reaction Rate of the Karl Fischer Titration Reaction. Part I. Potentiometric Measurements. J. Electroanal. Chem. (Lausanne) *71*, 305–315 (1976).
The reaction is first-order in water, sulphur dioxide and triiodide. For pH less than 5 the reaction-rate constant decreases logarithmically with decreasing pH.

7603 Lima, M.C.P.; Spiro, M.: Analysis of Organic Amines and Acids for Water. Anal. Chim. Acta *81*, 429–431 (1976).
The water content in 1,2-diaminoethane is titrated using a KF reagent based on 2-methoxyethanol. A mixture of acetic acid and 1,2-diaminoethane serves as the working medium.

7604 Rechenberg, W.: Determination of the Water Content of Cement. (In German.) Zement-Kalk-Gips *29*, 512–516 (1976).
The sample is heated in a stream of argon to 1000 °C. The argon is passed through an oxidation catalyst into the titration cell filled with anhydrous methanol.

7605 ISO 3699–1976 (E). Anhydrous Hydrogen Fluoride for Industrial Use – Determination of Water Content – Karl Fischer Method.
Addition of the sample to cooled, dry pyridine containing 5% anhydrous hydrogen fluoride. Titration with KF solution using a described apparatus.

7701 Farzaneh, A.; Troll, G.: Quantitative Method for Determining Hydroxyl and H_2O in Minerals, Rocks and Other Solids. (In German.) Fresenius Z. Anal. Chem. *287*, 43–45 (1977).
Samples are pyrolyzed in an induction furnace. The liberated water is continuously collected in a KF cell an titrated using a KF titrator.

7702 Bykova, L.N.; Petrov, S.I.; Khusainov, M.G.; et al.: Conditions for the Coulometric Determination of Small Amounts of Water in Aprotic Solvents. (In Russian.) Zh. Anal. Chim. *32*, 938–944 (1977) (Engl. summary).
Water contents of 0.05% are determined with high accuracy in dimethyl formamide, dimethyl acetamide and dimethyl sulphoxide. The reagent and working conditions are mentioned.

7703 Verhoef, J.C.; Barendrecht, E.: Mechanism and Reaction Rate of the Karl Fischer Titration Reaction. Part II. Rotating Ring-Disc Electrode Measurements. J. Electroanal. Chem. (Lausanne) *75*, 705–717 (1977).
The kinetics of the oxidation of the monomethyl sulphite ion is investigated. The results agree with those of [7602].

7704 Verhoef, J.C.; Barendrecht, E.: Mechanism and Reaction Rate of the Karl Fischer Titration Reaction. Part V. Analytical Implications. Anal. Chim. Acta *94*, 395–403 (1977).
A two-component reagent is recommended. A methanolic sodium acetate-sulphur dioxide solution serves as a solvent and an iodine solution in methanol as a titrant.

7705 Hiranuma Sangyo / Miyake, S.; Suto, T.: A Solvent for the Determination of Water by the Karl Fischer Method. Jpn. Kokai Tokkyo Koho JP 53/93095 (1977).
Solvent for the coulometric determination of water in organic acids and ketones consisting of acetonitrile and ethylene glycol and/or propylene glycol.

7707 Kreiser, W.R.; Martin, R.A.: Comparison of Accuracy, Precision and Speed of Three Methods for Determining Moisture in Milk Chocolate. J. Assoc. Off. Anal. Chem. *60*, 303–306 (1977).
Water content was determined by oven drying, azeotropic distillation and KF titration. The Karl Fischer titration is more accurate, more precise and much faster.

7708 Martin, R.A.: Collaborative Study of the Determination of Moisture in Chocolate by Karl Fischer Titration. J. Assoc. Off. Anal. Chem. *60*, 654–657 (1977).
Five samples were analysed by 10 collaborators. The Karl Fischer method gives better precision than the air-oven method.

7709 DIN 10766. Analysis of Coffee and Coffee Products; Determination of Water Content of Green Coffee. (In German.)
The grinded sample is extracted with dioxane for 20 min followed by azeotropic distillation. Methanol is added and the water content is titrated with KF solution.

7710 ISO 4317–1977 (E). Surface-Active Agents – Determination of Water – Karl Fischer Method.
Powders, pastes or solutions are added to methanol. The water content is titrated according to ISO 760 [7806].

7801 Zürcher, K.; Hadorn, H.: Determination of Water in Food Stuffs by the Karl Fischer Method. (In German.) Dtsch. Lebensm. Rundsch. *74*, 249–259 (1978).
Flour, margarine, corn starch and dried apricots are analysed by vacuum drying, air-oven drying, azeotropic distillation and KF titration. The KF titration at increased temperature is to be preferred.

7802 Cedergren, A.: Reaction Rates between Water and Some Rapidly-Reacting Karl Fischer Reagents. Talanta *25*, 229–232 (1978).
The KF reaction can be accelerated by adding formamide, dimethylformamide or N-methylformamide.

7803 Evenko, G.N.; Kats, B.M.; Sherman, F.B.: Micro Determination of Water in Weak Basic Anion Exchangers with Fischer Reagent. (In Russian.) Zh. Anal. Chim. *33*, 314–316 (1978) (Engl. summary).
The sample is extracted with methanol and titrated with KF solution. Alternatively the sample is extracted at 50 °C.

7804 Zürcher K.; Hadorn, H.: Determination of Water in Foodstuffs using the Karl Fischer Method. Part 2. (In German.) Dtsch. Lebensm. Rundsch. *74*, 287–296 (1978).
Different methods of pre-treatment of the samples are tested. Titrations at different temperatures, recording of titration curves, back-titration methods, extraction at different temperatures are applied to determine the water content. Part I see [7801].

7805 Mantovani, G.: Dry Substance in Sugar Products other than Sugar. Proc. 17th Sess. ICUMSA, Montreal 1978.
A collaborative study on the determination of dry substance in molasses and liquid sugar by KF titration and oven drying.

7806 ISO 760–1978 (E). Determination of Water – Karl Fischer Method (General Method).
Recommendations for the production of KF reagents using methanol or 2-methoxyethanol as a solvent. Two methods of titration are specified, the visual method and the electrometric method.

7807 DIN 10240 part 3. Analysis of Tobacco; Machine-Smoking of Cigarettes and Determination of Smoke Condensate; Determination of Crude Condensate and Nicotine-Free Dry Smoke Condensate.
This part deals with the water determination in smoke condensate using the Karl Fischer titration. The nicotine-free dry smoke condensate is calculated using the water content.

7808 ASTM D 1364–78. Standard Test Method for Water in Volatile Solvents (Fischer Reagent Titration Method).
The water content is titrated using pyridine-ethylene glycol as a working medium. This mixture is dried by pre-titration. Then the sample is added and analysed.

7809 ASTM Designation: D 1744–64 (Reapproved 1978). Standard Test Method for Water in Liquid Petroleum Products by Karl Fischer Reagent.
A mixture of methanol and chloroform (1 + 3) is used as a working medium. The mixture is dried by pre-titration. Then the sample is added to the pre-dried solvent and titrated immediately.

7810 Francis, H.J.; Persing, D.D.: Determination of Water in Thiols by Karl Fischer Titration. Talanta *25*, 282–283 (1978).
Water in thiol-containing samples may be determined by titration with Karl Fischer reagent after conversion of the thiol into a thioether. Both acrylonitrile and N-ethylmaleinimide have been found to be suitable.

7811 ASTM Designation: D 3277–73 (Reapproved 1978). Standard Test Method for Moisture Content of Oil-Impregnated Cellulosic Insulation.
The samples are extracted using a mixture of methanol and chloroform dried by pre-titration with KF solution. After extraction the water content is titrated in the same way.

7901 Seidelmann, U.: Determination of Traces of Water in Cyanogen Chloride by Karl Fischer Titration. (In German.) Fresenius Z. Anal. Chem. *295*, 380–381 (1979).
The cyanogen chloride is added to ethylene glycol and evaporated. The water retained by ethylene glycol is then titrated with Karl Fischer solution.

7902 Yamaya, K.: Determination of Water Content in Muscovites Containing Large Amounts of Fluorine by Karl Fischer Titration after Fusion with Silicon Dioxide. Anal. Chim. Acta *110*, 233–243 (1978).
Muscovite is mixed with silicon dioxide and heated to 1200 °C. The water liberated is carried by dried nitrogen into the KF cell containing methanol and ethylene glycol. Titration with KF solution.

7903 Zürcher, K.; Hadorn, H.: Working Instructions for the Determination of Water in Foodstuffs by the Karl Fischer Method. (In German.) Mitt. Gebiete Lebensm. Hyg. *70*, 485–496 (1979).
Water determination is only possible if the water is freely available. If it is en-

closed by the cells the sample has to be triturated or titrated at high temperatures.

7904 J.T. Baker Chemicals / Verbeek, A.E.; Mattheij, J.M.J.: Reagent for the Quantitative Determination of Water and its Application to Determine Water. (In German.) Patent application EP 00023 230, (1979).
Two-component reagent. The solvent-component consists of sulphur dioxide and an acetate dissolved in a mixture of methanol and 2-methoxyethanol.

7905 Wiese, G.; Mannaa, A.: On Dilatometrically Indicated Titrations. XVI. Determination of Water Content in Salts by Dilatometric Titrations in Organic Solvents. (In German.) Fresenius Z. Anal. Chem. *299*, 257–260 (1979).
The water content is analysed in 34 salts. The determination can be carried out with sufficient accuracy.

7906 ASTM D 1348–61 (Reapproved 1979). Standard Test Method for Moisture in Cellulose.
The methods cover the determination of moisture in cellulose using two oven-drying methods and one Karl Fischer procedure. Cellulose is extracted with methanol. An excess of KF solution is added to the extract followed by back-titration.

7907 ASTM Designation: E 700–79. Standard Test Method for Water in Gases using Karl Fischer Reagent.
The gas is transferred slowly through a scrubber liquid pre-titrated with KF solution to a stable end point. The absorbed water is titrated in the same way.

7908 ASTM Designation: D 1533–79. Standard Test Method for Water in Insulating Liquids (Karl Fischer Method).
A mixture of methanol and chloroform (1 + 2) is dried by pre-titration with the KF reagent. Then the sample is added and the water content is titrated.

7909 Christen, G.L.; Richardson, G.H.: Modified Karl Fischer and Vacuum Moisture Methods for Cheese Products: Comparative Study. J. Assoc. Off. Anal. Chem. *62*, 828–831 (1979).
Methanol-chloroform is used to extract moisture from cheese after homogenization using a Brinkmann Polytron PT 10 homogenizer.

8001 Kågevall, I.; Åström, O.; Cedergren, A.: Determination of Water by Flow-Injection Analysis with the Karl Fischer Reagent. Anal. Chim. Acta *114*, 199–208 (1980).
To analyse organic solvents the KF reaction is combined with a flow-injection system. 2 µl of the sample are injected into a stream of KF solution (1 ml/min).

8002 Scholz, E.: Karl Fischer Reagents without Pyridine. (In German.) Fresenius Z. Anal. Chem. *303*, 203–207 (1980).
Other amines than pyridine can be used to produce KF reagents of superior quality. They are characterized by stable end points and a high speed of reaction.

8003 Riedel-de Haën AG / Scholz, E.: Reagent for Titration and its Use. (In German.) DE 30 08 421 (1980).
Karl Fischer reagent consisting of sulphur dioxide, iodine, an aliphatic amine and an organic solvent. The amine can contain hydroxy groups. Mono-, di- and triethanolamine are mentioned.

8004 Merck Patent GmbH / Fischer, W.; Krenn, K.-D.: Pyridine-Free Karl

Fischer Reagent and Method for the Determination of Water using this Reagent. (In German.) Pat. Application DE 30 10 436 (1980).

Two-component reagent. The solvent component consists of sulphur dioxide and an amine. Mono-, di- and triethanolamine are mentioned.

8005 Mitsubishi Chemical Industries / Muroi, K. M.; Fujino, H. Y.: Karl Fischer Reagent. (In German.) Pat. application DE 30 40 474 (1980).

One-component reagent containing iodine, sulphur dioxide, pyridine and alkylene carbonate as a solvent.

8006 Sherman, F. B.: Determination of Water with a Modified Karl Fischer Reagent. Stability and the Mechanism of Reaction with Water. Talanta *27*, 1067–1072 (1980).

Using a DMF-based reagent, water is titrated in different organic solvents. The titre of the reagent depends on the solvent applied. In methanol 0.30 mg water per ml are found, in n-butanol 0.45 and in DMF about 0.57. The stoichiometric course of the reaction is discussed.

8007 Petrov, S. I.; Galitsyn, A. D.; Kasperovich, V. L.: New Electrolyte for Coulometric Determination of Water. (In Russian.) Zh. Anal. Chem. *35*, 2195–2198 (1980). (Engl. summary).

The Karl Fischer reagent contains 2-ethoxyethanol instead of methanol. It can be applied to the determination of water in carbonyl compounds like acetone.

8008 Hadorn, H.: Collaborative Study on Water Determination in Foodstuffs by Karl Fischer Titration. (In German.) Mitt. Gebiete Lebensm. Hyg. *71*, 220–235 (1980).

5 laboratories. Corn starch is titrated at room temperature. Corn starch and flower are titrated at increased temperature. Sultanas are triturated and analysed.

8010 Riedel-de Haën AG / Scholz, E.: Titration Agent and Method for using same. USP 4,378,972; EP 0035 066; DE 30 39 511 (1980).

Karl Fischer reagent consisting of iodine, sulphur dioxide and a nitrogen-containing heterocyclic compound having two hetero-atoms.

8101 Zürcher, K.; Hadorn H.: Water Determination According Karl Fischer in Different Foodstuffs. (In German.) Dtsch. Lebensm. Rundsch. *77*, 343–355 (1981).

Close to 100 foodstuffs were submitted to water determination by Karl Fischer and by conventional methods. Results were discussed. Working techniques are recommended for all products under investigation.

8102 Scholz, E.: Karl Fischer Reagents without Pyridine (3). The Accuracy of the Determination of Water. (In German.) Fresenius Z. Anal. Chem. *306*, 394–396 (1981).

Non-pyridinic reagents react more rapid und give more accurate results. Best results are obtained when using two-component reagents.

8103 Scholz, E.: Karl Fischer Reagents without Pyridine (4). One-Component Reagents. (In German.) Fresenius Z. Anal. Chem. *309*, 30–32 (1981).

KF reagents can be prepared using diethanolamine or morpholine instead of pyridine. The new reagents are better suited for water determination because the titration proceeds faster.

8104 Scholz, E.: Karl Fischer Reagents without Pyridine (5). New Substances for

Standardization. (In German.) Fresenius Z. Anal. Chem. *309*, 123–125 (1981).
Iso-butanol/xylene and similar mixtures of solvents are more suitable for the standardization of KF solutions than water-in-methanol standards normally used.

8105   Jones, F. E.: Determination of Water in Solids by Automatic Karl Fischer Titration. Anal. Chem. *53*, 1955–1957 (1981).
The sample and milling balls are put into a milling vessel. Methanol is added and the sample is milled for 15 min. After centrifuging a part of the methanol is titrated.

8106   Zürcher, K.; Hadorn, H.: Artifacts Encountered in Determining Water Content by means of the Karl Fischer Method. (In German.) Mitt. Gebiete Lebensm. Hyg. *72*, 177–182 (1981).
For fat-containing samples methanol/chloroform is used as a working medium. Sugar-containing samples may decompose at higher temperatures. Acids accelerate the decomposition. The addition of diethylamine is recommended.

8107   Jones, F. E.; Brickencamp, C. S.: Automatic Karl Fischer Titration of Moisture in Grain. J. Assoc. Off. Anal. Chem. *64*, 1277–1283 (1981).
5 g of grain, 50 ml of methanol and milling balls are put into the milling vessel. Milling time 15 min. After centrifuging a part of the methanol is titrated. Detailed description of the procedure.

8108   Fujino, H.; Muroi, K.; Morishita, S.: Determination of Water Content in Electric Insulating Papers by the Karl Fischer Method. (In Japanese.) Bunseki Kagaku *30*, 624 (1981) (Engl. summary).
Nitrogen-carrier extraction at higher temperatures followed by KF titration, methanol extraction and oven drying are compared. Method 1 is most rapid and most accurate.

8109   ASTM Designation: D 890–58 (Reapproved 1981). Standard Test Method for Water in Liquid Naval Stores.
Liquid samples are analysed using a two-component reagent. The solvent-component can be diluted with benzene and methanol.

8110   ASTM Designation: D 789–81. Standard Specification for Nylon Injection Molding and Extrusion Materials.
To determine moisture the sample is heated to 180–260 °C. The water evaporated is condensed in a cold trap and titrated using a diluted KF reagent.

8111   Riedel-de Haën AG / Scholz, E.: Two-Component Karl Fischer Reagent. DE 31 36 942.
Karl Fischer reagent with a titrant component containing a non-hygroscopic organic solvent.

8112   ASTM Designation: E 203–75 (Reapproved 1981). Standard Test Method for Water Using Karl Fischer Reagent.
A general guide for the application of the KF method for determining water in solid or liquid organic and inorganic compounds. The solvent methanol is pre-titrated with the KF solution then the sample is added and titrated. A one-component reagent is described. Informations on standardization and on interferences.

8113   ASTM Designation: D 4017–81. Standard Test Method for Water in Paints and Paint Materials by Karl Fischer Method.

The material is dissolved in pyridine, or another solvent, and titrated directly with the KF reagent. Pyridine is used as a solvent because methanol will not dissolve resins and to minimize problems caused by ketones.

8114 Schneider, F. H.: Water Determination in Raw Materials of Vegetable Oils. (In German.) Fette Seifen Anstrichm. *83*, 329–337 (1981).
The reliability of drying methods is discussed. Using the cabinet drying method resorption of water during cooling has to be prevented. The radiation drying is disturbed by uneven temperature distribution. A drying method using 'hot weighing', is presented.

8115 Dahms, H.: Karl Fischer Reagent and its Use. USP 4,354,853 (1981).
A two-component reagent is claimed, using a titrant component with a solvent having a low solubility for water.

8201 Scholz, E.: Karl Fischer Reagents without Pyridine (6). Determination of Water in Carboxylic Acids. (In German.) Fresenius Z. Anal. Chem. *310*, 423–426 (1982).
About 50 acids are titrated volumetrically. Most of the acids can be analysed without interferences. Formic acid tends to esterification.

8202 Koupparis, M. A.; Malmstadt, H. V.: Determination of Water by an Automated Stopped-Flow Analyzer with Pyridine-Free Two-Component Karl Fischer Reagent. Anal. Chem. *54*, 1914–1917 (1982).
0.5 ml samples are mixed with 2 ml Karl Fischer solvent by injection in glass vials. After addition of the titrant the absorption is measured photometrically.

8203 Scholz, E.: Karl Fischer Reagents without Pyridine (7). Two-Component Reagents Containing Imidazole. (In German.) Fresenius Z. Anal. Chem. *312*, 462–464 (1982).
Six different reagents were prepared. All reagents give a stoichiometric reaction with water but a different speed of reaction.

8204 Weintraub, R.; Apelblat, A.: Determination of Water in Hydrogen Bromide Gas. Anal. Chem. *54*, 2627–2628 (1982).
HBr is allowed to flow into a cell containing pyridine. The cell is cooled by an ice-water bath and the water content is titrated.

8205 ISO/DIS 6296. Liquid Petroleum Products – Determination of Water – Karl Fischer Method.
Water determination according to ISO 760 [7806]. 50 ml of a solvent selected on the basis of experience are dried by pre-titration. Then the sample is added and titrated as usual.

8206 ISO/DIS 5381. Starch Hydrolysis Products – Determination of Water Content – Modified Karl Fischer Method.
A mixture of methanol and formamide (70 + 30) is pre-titrated. The sample is suspended in this solvent and titrated with the KF reagent.

8207 'Schweizerisches Lebensmittelbuch', Chapter 22, Dietician Foodstuffs. (In German.)
This chapter is newly edited. It includes the KF titration to determine the water content in foodstuffs (5 pages). Equipment, reagents, titration technique and treatment of the samples are described.

8208 Dirscherl, A.: Micro Determination of Water in Carbonyl Compounds by

Means of a New K. Fischer Reagent. (In German.) Mikrochim. Acta (Wien) 1982, II, 477–484.

5 ml of methanol or methanol-chloroform are used as a working medium. 0.1 ml-samples are titrated at room temperature or in an ice bath.

8301 Scholz, E.: Karl Fischer Reagents without Pyridine (8). Coulometric Determination of Water. Fresenius Z. Anal. Chem. *314*, 567–571 (1983).

Pyridine-free KF reagents are suitable for the coulometric determination of water. An accelerated reaction improves the accuracy. 100 substances are tested in three different coulometers.

8302 Scholz, E.: Determination of Water in Foodstuffs. Karl Fischer Reagents without Pyridine (9). (In German.) Dtsch. Lebensm. Rundsch. *79*, 302–306 (1983).

The new reagents accelerate the reaction and a determination of water takes normally 2–10 min. If the water has to be extracted the sample should be titrated at higher temperatures.

8303 DIN 51777 part 1. Testing of Mineral Oil Hydrocarbons and Solvents; Determination of Water Content According to Karl Fischer; Direct Method. (In German.)

The products soluble in 2-methoxyethanol or in methanol/propanol are titrated directly using these alcohols as a working medium. Pyridine-free reagents are to be preferred.

8304 Muroi, K.; Fujino, H.: Determination of Water in Ketones and Silicon Oils by Karl Fischer Titration. Sekiyu Gakkaishi *26*, 97–101 (1983), CA 98(22):191007p.

The reagent contains chloroform, propylene carbonate, pyridine and sulphur dioxide. Similar compositions for volumetric and coulometric application. For volumetric determinations alcohol-free titrants are required.

8305 ISO/DIS 7105. Liquefied Anhydrous Ammonia for Industrial Use – Determination of Water Content – Karl Fischer Method.

Evaporation of a test portion in presence of ethanediol and determination of the water content of the residue by the Karl Fischer direct electrometric method.

Author Index

Italicized numbers refer to publications listed on pages 111–131.

van Acker, P. 86, 108, *7303, 7404*
Allen, G.A. 34, *4902*
Andrysiak, A. 76, *7201*
Andrysiak, E. 76, *7201*
Apelblat, A. 86, *8204*
Archer, E.E. 61, 105, *6501, 7405*
Ashby, E.C. 90, *4101, 4102*
Åström, O. 23, *8001*

Barbi, G. 86, *6305*
Barendrecht, E. 4, 6, 7, 10, 22, 24, 60, *7602, 7703, 7704*
Barger, J.D. 107, *7005*
Barnes, L. 76, *5902*
Becker, H.C. *4903*
Belin, P. *7002*
Below, L. 105, *7503*
Beyer, H. 77, *6601*
Birett, 71
Bizot, J. 32, 60, *6702*
Blasius, E. 4, *5302*
Blomgren, E. 36, 37, *5201, 5702*
Bolling, H. 22, 103, *5803*
Bonauguri, E. 5, 42, *5502*
Bonderovskaja, E.A. 107, *7501*
Boyd, C.M. 19, 32, 54, *5901*
Brabson, J.A. 109, *6909*
Brickenkamp, C.S. 103, *8107*
Brobst, K.M. *4801*
Bryan, W.P. *7601*
Bryant, W.M.D. 3, 5, 27, 30, 35, 37, 39, 77, 82, 90, 91, *3901, 4001, 4101, 4102*
Burns, E.A. 90, *6201, 6306*
Bykova, L.N. 83, *7702*

Čadersky, I. 92, *7203*
di Caprio, B.R. 88, *4702*
Caro, J.H. 109, *6405*
Cedergren, A. 4, 6, 7, 23, 32, 37, *7401, 7402, 7403, 7802, 8001*

Chiang, P.T. 89, *6304*
Christen, G.L. 101, *7909*
Cooper, W.C. 89, *6304*
de Corte, F. 86, 108, *7303, 7404*
Cserfalvi, T. 108, *7305*
Cunningham 80
Czichon, P. *7001*

Dahms, H. *7204, 8115*
Davies, R.J. 105, *7502*
Delfs, E.-M. 98, 101, *7202*
Delmonte, C.S. *7004*
Dietrich, K. 107, *6708*
Dinse, H.-D. 107, *6908, 7007*
Dirscherl, A. 78, *8208*
Duncan, R.D. 109, *6909*

Eberius, E. 1, 2, 34, 35, 36, 37, 39, 42, 77, 94, 99, 111, *5603, 5703*
Emmerich, A. 98, *7102, 7505*
Epps, E.A. 94, 98, *6602*
Evenko, G.N. 108, *7803*

Farzaneh, A. 110, *7701*
Fischer, F. 77, *6401*
Fischer, K. 1, 2, 3, 5, 15, 27, 34, 38, 42, 88, *3501*
Fischer, W. *8004*
Fligier, J. *7001*
Francis, H.J. 83, *7810*
Fujino, H. 77, 108, *8005, 8108, 8304*

Galitsyn, A.D. 78, *8007*
Gimesi, O. 108, *7305*
Giuffre, L. 86, *6402*
Glöckner, G. 107, *5903*
Glover, J.H. 90, *5203*

Hadorn, H. 22, 94–97, 99–104, *7801, 7804, 7903, 8008, 8101, 8106*
Heumann, W.R. 108, *6605*
Hibbits, J.O. 110, *5401*

Hilton, J. 105, *7405*
Hodgson, H.W. 90, *5203*
Holden, T.F. 101, *6802*
Hopfinger, A. *7001*
Hopkinson, F.J. 25, 29, 77, *4301*
Hoste, J. 108, *7303, 7404*

Ishii, Y. 74, 77, 105, 107, *6502, 6604, 6701, 6801*

Jeater, H.W. 61, *6501*
Jecko, G. 110, *7504*
Jenner, H. 36, 37, *5201, 5702*
Johansson, A. 17, 28, 30, 79, 81, 89, *4701, 5601*
Johnson, C.M. 94, 97, 99, 103, *4501*
Johnston, I. 109, *6404*
Jones, F.E. 96, 103, *8105, 8107*
Jungnickel, J.L. 28, 34, 77, *5501*

Kågevall, I. 23, *8001*
Kasperovich, V.L. 78, *8007*
Kats, B.M. 108, *7803*
Kawinski, F. 8, 107, *6102*
Kellum, G.E. 92, 107, *6603, 6705, 6706, 7005*
Kempf, W. 94, 99, *5703, 7205*
Keyworth, D.A. 21, 32, 60, *6301*
Khusainow, M.G. *7702*
Kindsvater, H.M. *5602*
Klimova, V.A. 5, 29, 30, 60, 77, *6703, 6704, 7302*
Knight, W.P. *5602*
Koizumi, H. 97, *6906*
Kolthoff 25
Koupparis, M.A. 23, *8202*
Kowalski, W. 42, *5603*
Kreiser, W.R. 102, *7707*
Krenn, K.-D. *8004*
Kropotova, E.D. 107, *7501*
Kühn 71

Kviesitis, B. 99, *7304*

Laurene, A.H. *5202*
Levin, H. *4903*
Levy, G.B. *6904, 7204*
Lima, M.C.P. 81, *7603*
Linch, A.L. *5102*
Lindner, B. 110, *6905*
Lugowska, M. 83, *7001*
L'vov, A.M. 5, 29, 60, 77, *6703, 6704*

Majewska, J. *6503*
Malmstadt, H.V. 23, *8202*
Mannaa, A. *7905*
Mantovani, G. *7805*
Martin, R.A. 94, 95, 102, *7707, 7708*
Masko, T.E. 107, *7501*
Matheij, J.M.J. *7904*
McComas, W.H. 34, *4902*
McComb, E.A. 97, *5701*
van der Meulen, J.H. 30, *5101, 5301*
Meuser, F. *7205*
Meyer, A.S. 19, 32, 54, *5901*
Meyer, W. 107, *5903*
Mika, V. 92, *7203*
Milberger, E.C. 86, *4903*
Mitchell, J. 3, 4, 5, 27, 29, 30, 32, 35, 37, 38, 77, 82, 90, 91, 94, 108, *3901, 4001, 4002, 4101, 4102*
Miyake, S. 32, 82, *7406, 7408, 7705*
Moberg, M.L. 90, *5602*
della Monica, E.S. 101, *6802*
Morishita, S. 108, *8108*
Mrowetz, G. 98, 101, *7202*
Muraca, R.F. 90, *6201, 6306*
Muroi, K. 16, 63, 74, 77, 80, 86, 97, 99, 105, 107, 108, *6101, 6202, 6302, 6303, 6502, 6605, 6701, 6801, 6906, 7003, 7101, 7301, 8005, 8108, 8304*

Nair, J.H. *4802*

Oehme, F. 32, 54, *5801*
Ogawa, K. 16, 63, 76, 77, 107, *6302, 6303, 6502, 6604, 6701*
Ogorzalek, A. *6503*
Ono, M. 74, 80, 86, 105, *6801, 7003, 7101, 7301*

Pawlak, M.S. 76, *5902*
Persing, D.D. 83, *7810*
Peters, E.D. 28, 34, 77, *5501*
Petrov, S.I. 78, *7702, 8007*
Petrova, T.N. *7702*
Pizzini, S. 86, *6305*
Pollio, F.X. 108, *6307*
Praeger, K. 107, *6908, 7007*
Pungor, E. 108, *7305*

Rader, B.R. 104, *6707*
Rao, P.B. *7601*
Ravaine, D. 110, *7504*
Rechenberg, W. 89, 110, *7604*
Reid, V.W. 107, *6103*
Reinisch, G. 107, *6708*
Richardson, G.H. 101, *7909*
Rochon, F.D. 108, *6605*
Rudert, V. 110, *6905*

Sailly, M. *7002*
Sandell, D. 94, 99, *6001*
Schiene, R. 77, *6401*
Schneider, F. 98, *7102, 7505*
Schneider, F.H. 93, *8114*
Scholz, E. *8002, 8003, 8010, 8102, 8104, 8111, 8201, 8203, 8301, 8302*
Schroeder, C.W. *4802*
Seaman, W.M. 34, *4902*
Seibel, W. 22, 103, *5802, 5803*
Seidelmann, U. *7901*
Seltzer, D.M. *6904, 7204*
Seniga, G. 5, 42, *5502*
Sharma, H.D. 108, *6907, 7006*
Sherman, F.B. 5, 29, 30, 60, 77, 108, *6703, 6704, 7302, 7803, 8006*
Sistig, E. 24, 105, *7103*
Smith, D.M. 3, 4, 5, 27, 29, 30, 32, 35, 37, 38, 77, 82, *3901, 4001, 4101, 4102*
Smith, E.J. 109, *6404*
Smith, R.C. 92, *6603, 6705, 6706*
Spiro, M. 81, *7603*
Strange, T.E. 101, *6901, 7008, 7206*
Subramanian, N. 108, *6907, 7006*
Suto, T. 32, 82, *7408, 7705*
Swensen, R.F. 21, 32, 60, *6301*

Thomasow, J. 98, 101, *7202*
Thung, S.B. 103, *6403*
Ticmanis, U. 98, *7102, 7505*
Troll, G. 110, *7701*
Tsutsumi, C. 97, *6906*
Turner, L. 107, *6103*

Uhrig, K. *4903*

Varga, K. 77, *6601*
Verbeek, A.E. *7904*
Verhoef, J.C. 4–7, 10, 22, 24, 60, *7602, 7703, 7704*

Wachtel, F. 4, *5302*
Weintraub, R. 86, *8204*
Wernimont, G. 25, 29, 77, *4301*
Wiese, G. *7905*

Yamaya, K. 110, *7902*

Zabokritskij, M.P. 30, *7302*
Zucker, D. 110, *5401*
Zürcher, K. 22, 94–97, 99–104, *7801, 7804, 7903, 8101, 8106*

Subject Index

Accuracy 22, 31, 40, 95, 100
Acetaldehyde 76, 77
Acetals 76–78
Acetic acid 30, 32, 59, 61, 79, 80, 81
Acetone 12, 13, 76, 77
Acetonitrile 32, 82
Acid-base-titrations 7–13
Acids 58, 60, 61, 62, 78–80
Acrylic acid esters 80
Acrylonitrile 83
Air 69
Air humidity see moisture, atmospheric
Alcohol 75, 101
Aldehyde 23, 29, 62, 76–78
Alkali alcoholate 75
Alkali silicate 92
Alkaloides 82
Alkohol-water-mixture 42
Alkyl carbonate 13, 29
Almonds 102
Amide 82
Amidosulphuric acid 90
Amine 61, 62, 79, 80–82
Ammonia 90
Aniline 3, 30, 35, 36, 38, 81
Anodic oxidation 19
Anolyte 19, 20, 21, 31–33
Antimony compounds 91
Apricots 94
Aquamicron 33
Arsenic compounds 91
Ascorbic acid 62, 79, 96
Azobenzene 32

Baby food 102
Back titration 19, 21–22, 49, 50, 73, 96, 99
Bases, aliphatic 30
Bases, heterocyclic 31, 36
Bases, stabilizing 36, 61
Benzene 5, 32, 59
Benzoic acid 61, 81
Beverages 101
Biamperometric indication 25, 66
Bicarbonate 91
Bipotentiometric indication 25, 65, 66

Biscuits 103
Bisulfite addition 77
Blank titration 16, 75
Blank value 16, 23, 98
Boron compounds 92
Bread 103
Buffer 4, 7, 61, 65, 79
Bunsen reaction 3, 12, 31, 40, 42
Butanol 12, 36, 43, 59, 77, 101
Butter 100

Calcium sulphate 89
Candy 99
Caramel 97, 99
Carbohydrates 94, 97–100
Carbonate 62, 87, 91
Catholyte 20, 31–33
Cellulose 94, 108
Cement 109
Cheese 101
Chloral 76
Chlorides, inorganic 85–86
Chlorine, free 74, 75
Chloroform 16, 19, 59, 60, 72, 73, 74, 77, 85, 98, 100, 101, 106, 107
Chloromethane 74
Chocolate 94, 95, 102
Cigarette smoke 109
Citric acid 41, 79
Cocoa 102
Coffee 102
Coffee beans 96
Continuous determination 23–24
Corn 103
Cotton 108
Coulometric equipment 54–56
Coulometric titration 19–21, 24, 31–33, 70, 74, 75, 81, 105, 106, 107
Crude oil 106
Cupric chloride 86
Cupric sulphate 89
Curd 101
Cyanohydrine 77, 82
Cyclohexanone 76

Dead-stop indication 25
Dextrin 99
1,2-Diaminoethan 81
Dichloroacetate 4
Diethanolamine 7, 35, 36, 37, 65
Diisopropyl ketone 76
Dimethylaniline 30, 35
Dimethyl ether 76
Dimethylformamide 5, 29, 32, 60, 77, 82
Dimethyl sulphoxide 5, 83
Dipentene 109
Distillation, azeotropic 86, 94, 96, 102, 106, 109
Disulphides 83
Dithiocarbamate 83
Dithionite 62, 89
Drying of air 45–46, 52, 63–64

Egg powder 102
Electrodes, polarized 25
End point cut-off 49, 52, 53, 54, 66, 78
End point stability 16, 18, 21, 24, 31, 33, 48, 65, 66
End point titrator 48–49
Equipment 45–56
Error, sources of 57–70
Esterification 58, 79, 80, 86, 88, 92
Esters 78–80
Ethanolamine 30
Ethers 76
Ethyl acetate 5, 59
Ethylene glycol 59, 69, 90, 105, 106, 109
Ethylene oxide 76
N-Ethylpiperidine 61
Evaporation technique 21, 107, 108
Extraction of moisture 22, 49, 68, 96, 97, 99, 100, 103, 106

Fats 100
Ferric chloride 86
Fertilizers 109
Flour 94, 99–100
Flow-injection analysis 23
Fluorides 86
Foodstuffs 22, 93–104
Formaldehyde 76, 77
Formamide 5, 13, 32, 60, 72, 82, 85, 96, 98, 99, 101, 102, 103
Fructose 97, 98
Fruit, dried 103–104
Fruit juices 101

Galactose 97, 98
Gases 20, 50, 62, 69–70, 74, 105
Gases, liquified 70
Gelatine 100
Glucose 97

Glycerol 75
Grain 103
Grease 107
Groats 100

Halides, inorganic 85–86
Hazelnuts 102
Heterocycles 31, 80
Honey 99
Hydranal 17, 30, 31, 33, 43, 48, 78, 97, 98, 99, 100, 101, 102, 103
Hydrates for standardization 41
Hydrates of salts 41, 58, 67, 68, 85, 86, 88, 97, 109
Hydrazine 90
Hydrazine derivates 82
Hydrobromic acid 86
Hydrocarbons 69, 73–74
Hydrocarbons, halogenated 74–75
Hydrochloric acid 86, 105
Hydrofluoric acid 86
Hydrogen 105
Hydrogen peroxide 87
Hydroxamic acid 82
Hydroxides 62, 87–88
Hydroxylamine 90
Hypophosphite 91

Imidazole 10, 31, 35, 36, 37, 65, 79, 80
Indication 24–25, 47–48, 65–66
Inorganic compounds 85–92
Instant coffee 102
Insulating oil 107
Insulating paper 108
Iodopyridine 35, 37
Ion exchange resins 108
Iron ore 110
Isopropylamine 37

Joghurt 101
Juices 101

Ketals 76–78
Keto-acids 80
Ketones 23, 29, 32, 62, 76–78, 109
Kick-off indication 25
Kinetics 6

Lactose 97, 98
Lead oxide 87
Liquids, handling of 68
Loss on drying 94, 99
Lutidine 38, 39

Maltose 97, 98
Margarine 94, 100
Marzipan 103

Subject Index

Mayonnaise 101
Meat 104
Mercaptans 62, 82–83
Methanol 10, 12
2-Methoxyethanol 28, 30, 34, 35, 36, 39, 40, 59, 62, 71, 77, 78, 91, 106
Methylene chloride 70
Methyl sulphite 8, 10, 13, 31
Methyl sulphuric acid 4, 7, 61
Methyl sulphurous acid 6, 9, 10, 13, 62
Milk 101
Milk powder 101
Milling of samples 96
Mineral oil 106
Minerals 110
Moisture, adherent 68, 80, 85, 94, 98, 109
Moisture, atmospheric 20, 21, 22, 25, 34, 43, 45, 48, 55, 63–65, 67, 68
Moisture, cellular 94
Molasses 94, 98
Muscovite 110

Natural products 105–110
Nitrate 90
Nitric acid 90
Nitriles 82
Nitrite 90
Nitro-compounds 82
Nitrogen 69, 105
Nitrogen compounds 80–82, 90
Nitroso-compounds 82
Noodles 103
Nylon 107, 108

Oils, edible 100
Oils, lubricating 107
Oils, mineral 106
One-component reagent 15, 28, 29, 30, 31
Organic Compounds 73–83
Organosilicon compounds 92
Oven drying 94, 98, 99, 101, 107, 108
Overtitration 18, 65, 66
Oxalic acid 41, 79
Oxides 62, 87–88
Oxidizing substances 62
Oxygen 69, 105

Paints 109
Paper 108
Particle size 95, 99, 100
Pasta 103
Pastries 103
Pectin 100
Perchloric acid 7–13
Peroxides 87–88
pH 6, 24, 31, 36, 60–61, 65, 79, 80
Phenols 75

1,2-Phenylenediamine 81
Picoline 38, 39, 80
Pinene 109
Pipettes 63, 68
Phosphates 90–91
Phosphoric acid 90
Phosphorus compounds 90–91
Plastics 107–108
Polyamide 107, 108
Polycaprolactum 107
Polycarbonate 107
Polyethers 76
Polyethylene 107
Polyethylene terephthalate 107, 108
Polymethacrylate 107
Polypropylene 107
Polystyrene resins 108
Potentiograph 48–49
Pre-titration 15, 17, 41, 61, 66, 69, 75, 79, 80
Primary standard 41
Propanol 10, 11, 12, 19, 36, 59, 73, 74, 101, 106
Propellants 105
Propionic acid 61, 75, 79, 80
Propylene carbonate 35, 77
Propylene oxide 76
Protein foods 101
Pudding powder 102
Pyridine as a solvent 16, 60, 69, 77, 78, 105, 107
Pyrrol 81

Quinoline 30, 80

Reaction equation 3–13
Reaction, order of the 4
Reaction rate 6, 7, 13, 32, 48, 60
Reaction rate constant 6, 7, 22
Reagent, analysis of the 39–40
Reagent, low in methanol 29
Reagent, preparation of 38–40
Reagent, stability of 34–38
Reagent, stabilized 28
Reagent solutions 27–43
Reaquant 17, 30
Reducing substances 62
Refrigerants 105
Rice 103
Rusks 103
Rye 103

Saccharose 98
Safety precautions 70–72
Salicylate 4, 30
Salicylic acid 81
Salts 80, 85
Samples, handling of 66–70, 95–96, 106

Subject Index

Selenium compounds 89
Shelf life 18, 27, 28, 30, 31
Side reaction 21, 38, 59, 62
Silica gel 92
Silicon compounds 92
Sodium acetate 10, 30, 41, 65, 80
Sodium tartrate 41, 80
Solids, handling of 68
Stability of reagents 34–38
Standard techniques 15–25
Standardization 40–43
Standards, methanolic 42
Standards, nonhygroscopic 43
Stannous salts 62
Starch 94, 99
Stearyl alcohol 75
Stoichiometry 3–5, 7–10, 12, 17, 22, 29, 31, 58, 59, 60, 80
Stopped-flow analysis 23
Successive titration 17, 19
Sugar 22, 94, 97–99
Sulphate 88–89
Sulphide 62, 83
Sulphite 87, 89
Sulphonic acid 83
Sulphur compounds 82–83, 88–89
Sulphur dioxide, liquid 38, 39, 88
Sulphur trioxide 4
Sulphuric acid 87
Sulphurous acid 11
Sultanas 104
Surfactants 109
Syringes 16, 20, 41, 42, 68, 106

Technical products 104-110
Tellurium compounds 89
Temperature, titrating at different 22–23, 81, 96, 97, 99, 102, 103

Tetraethyl lead 106
Thiols 83
Thiosulphate 89
Thiourea 83
Titrant, diluted 70, 74
Titration techniques 15–25, 96–97
Titration vessel 15, 46–47, 50–54, 63, 64, 97, 105
Titre, decrease of 27, 28, 34–38, 39
Titre, standardization of 40–43
Toluene 5
Tributylamine 30
Triethanolamine 30
Two-component reagent 15, 17, 28, 30, 31, 40, 81

Uranium oxide 110
Urea derivates 82

Varnishes 109
Vegetables 94, 96, 103–104

Water as a primary standard 41–42
Water capacity 19
Water equivalent 16, 27, 28, 38, 39, 41, 63, 106
Water-in-methanol mixtures 42
Water of hydration 41, 58, 67, 85, 86, 88, 97, 109
Wax 81
Wet milling 96, 103
Wheat 103
Whey powder 101
Wood 108
Working medium 16, 49, 58–60, 73, 74

Xanthogenates 83
Xylene 5, 43, 59, 67